# PLANTATION FORESTRY IN THE AMAZON:

# THE JARI EXPERIENCE

by

Clayton E. Posey

Robert J. Gilvary

John C. Welker

L. N. Thompson

Edited by

Harold K. Steen

Forest History Society

Durham, North Carolina

1997

Printed in the United States of America

The Forest History Society is a nonprofit, educational institution dedicated to the advancement of historical understanding of human interaction with the forest environment. It was established in 1946. Interpretations and conclusions in FHS publications are those of the authors; the institution takes responsibility for the selection of topics, the competence of the authors, and their freedom of inquiry.

Work on this book and its publication were supported by grants to the Forest History Society.

Cover photo: Gmelina Plantation.
Courtesy of John C. Welker.

Library of Congress Cataloging-in-Publication Data

Plantation forestry in the Amazon : the Jari  experience / by Clayton E. Posey
... [et al.] ; edited by Harold K. Steen.
        p.   cm.
    Includes bibliographical references.
    ISBN 0-89030-054-2 (pbk.)
    1. Projecto Jari—History. 2. Investments, American—Amazon River
Valley. 3. Forests and forestry—Economic aspects—Amazon River Valley.
4.  Tree farms—Amazon River Valley—History—20th Century. 5.
Plantation workers—Amazon River Valley—Interviews.     I. Posey,
Clayton E.,   1938-          ,  II. Steen, Harold K.
HG5332.P59    1997
333.75'0981'1—dc21
                                                                97-36526
                                                                CIP

# CONTENTS

## Clayton E. Posey

# Robert J. Gilvary

# John C. Welker

# Lawrence N. "Tommy" Thompson

# Introduction

Daniel K. Ludwig's mammoth forestry project in Brazil captured many headlines in the 1970s and 1980s, and an occasional headline even today. Readers of financial, forestry, and environmentalist outlets were given a series of accounts, and the large majority of this writing has criticized the effort. In fact, everything about the story is large, with acres in the millions and dollars in the billions.

Even a casual reading of this material today shows that often the later authors used earlier authors as their main source. Anger is a common tone, as authors find fault with the science upon which the project depends, its financial structure and viability, and even Ludwig the man. Also, nearly all agree that the project was a failure. To the newcomer it all seems a bit strange; how could something be so universally bad?

Ludwig died on August 27, 1992 at age 95, taking parts of the story with him, so we will never really know. Nonetheless, the following recollections by four key participants add to the bits and pieces of the larger written record. So read on; it is a great story.

Long in the shipping business, Ludwig became immensely wealthy following World War II through building a fleet of supertankers; eventually he would own three hundred companies worldwide. A visionary with stout entrepreneurial spirit, he liked the challenge of investing in extractive industries, and on occasion an agricultural venture. He came to believe that there would be a shortage of fiber by the 1980s, which led him into his largest development project, this time along the Jari River in northern Brazil.

Ludwig had sent an aide, Everett Wyncoop, to find the most suitable tree species. Perhaps it was because he was a mining engineer, but Wyncoop eventually located *Gmelina arborea* (generally called gmelina) used as pit props in a Nigerian mine. The British had earlier transplanted the exceptionally fast-growing species from Burma. When he reported back to Ludwig that gmelina could produce fiber on a commercial basis in only six years, the shipping magnate authorized field testing. It grew well in Costa Rica, too, and Ludwig made the fateful decision to focus on the potential of that single species.

With vigorous gmelina seed collecting underway, the next step was to locate a large tract of land on deep water and in a country run by a development-friendly government. Political unrest in likely African countries prompted

Ludwig to look to the tropical portions of the western hemisphere. In 1964 he met with Roberto Campos, Brazilian minister of planning, and the way was cleared to proceed in the Amazon basin.

During this period of Brazilian history, about fifty million acres had been sold to multinationals. Thus, when Ludwig purchased about three million acres along the Jari River, a major tributary to the lower Amazon, it was not a unique transaction. Eventually he would clear a bit less than ten percent of the area and install plantations of fast-growing exotic species. Rudolfo Dourado was the engineer who headed up the initial clearing operation and Monte Dourado, the town that grew along with the project, was named for him.

Parallel to the development of the Amazon plantation, Ludwig had long been intrigued with the idea of constructing a processing facility under First World conditions and floating it into position at a Third World site. His concept was not limited to wood processing, but the timing was right. He would construct a full sized pulpmill and power generating plant in Japan, and tow them across the Indian and Atlantic oceans, and onto awaiting foundations downstream from Monte Dourado at the port of Munguba. The resultant pulp was successfully sold on the world market, but complications led to selling the enterprise to a consortium of Brazilian interests in 1982. What follows is related to the American portion of the story.

Our narrators consist of three foresters and an engineer, and their recollections are presented in the order that they were recorded. We begin with Clayton Posey, a Ph.D. forest geneticist who was to develop and manage the nurseries that produced the seedlings that would be used to create the plantations. He quickly moved to positions of ever-greater responsibility. The interview took place in Wayne, Oklahoma.

Next is Robert Gilvary, a civil engineer who arrived in Jari even before Posey. It was for him to design and construct towns, ports, railroads, truck roads, power lines, and the foundation for the gigantic floating mill. As he recounts, he grew into the extraordinarily challenging job and literally built Jari. The interview took place in Blacksburg, Virginia.

John Welker was trained in forestry and economics, and hired on just out of school. He found himself in charge of planning; establishing short- and long-term schedules so that there was overall coordination between the complex forestry pieces. After all, this was not an academic exercise; the trees that the plantations produced needed to be synchronized with the pulpmill. The interview took place in Columbus, Georgia, and just across the Chattahoochee River in Phenix City, Alabama.

"Tommy" Thompson is a forester with a great breadth of hardwood experience, including early work in the Amazon. He was one of many consultants that Ludwig retained to be his eyes and ears at Jari. However, he stands out from the rest in that he was the only consultant also used by Posey. Thus, among other talents, he was a useful intermediary between New York and Monte Dourado, many miles away and cultures apart. The interview took place in Milledgeville, Georgia.

Another name that appears frequently throughout is that of Johan Zweede, vice president for forestry operations at Jari. Zweede is still active in the Amazon and was not available to be interviewed.

Many of the questions asked were repeated for each interview; however, the answers were sometimes different. The differences stem from different vantage points and responsibilities, and often due to different timeframes—1970 looked rather different from 1980. Presumably, if the four interviewees could be brought to the table, most apparent differences would be easily ironed out.

The edited transcripts are of an oral record, and the way we talk is not the way we write. The conversational tone remains, at times with syntax and grammar of a sort that our composition teachers of an earlier time would have circled in red. Although many of the sections recreate conversations, quotation marks are rare.

The illustrations were provided by Tommy Thompson, John Welker, and Ken McNabb, who was not interviewed but who also had worked at Jari toward the end of the American phase. Over eighty Jari slides are held by the Forest History Society. Also a substantial body of research data is held by the Manuscripts Division, University Library, SUNY Syracuse.

The interviews and this publication were made possible through the generous support of T & S Hardwoods, Anderson-Tully Company, and Mr. and Mrs. Stanley R. Day.

Harold K. Steen

Durham, N.C.

Caribbean pine nursery. Welker Photo.

# CLAYTON E. POSEY

## Introduction

In the pages that follow, Clayton Posey gives an account that is at variance
with every report examined in preparation for the interview. He could offer
no rationale for the hostile treatment at the hands of a wide range of writers.
To him, the project was a success, both scientifically and financially. As you
will read, he speaks with pride and good humor about his years in association
with Daniel Ludwig.

Clayton earned a Ph.D. in forest genetics from North Carolina State
University and subsequently taught at Auburn University and Oklahoma
State University. His very successful professorial research caught the eye of a
Ludwig advisor, who saw the need for a geneticist to be on site in Brazil.
Thus, in 1969 the Posey family left Stillwater for a rather primitive situation
on the Jari River, a major tributary to the lower Amazon.

He quickly moved up to the manager of the Forestry Division, then to acting
executive director, and from 1972 to 1975 he was the chief executive officer.
Then Ludwig called him to New York to be vice president of the Forest
Products Division, Universe Tankerships, Inc. It is from these various
vantage points that Clayton reports what happened and why.

# The Initial Situation at Jari

Clayton E. Posey (CEP): The Jari project started in '67, and unfortunately Ludwig was a builder. He was not a biologist, and his concept of building something was to bring in construction people. And he wanted to build a forest. The way you build a forest is with "x" number of million of dollars' worth of construction equipment and "x" thousands of construction people who build dams and bridges and hotels and cities and whatever.

Of course, it didn't work. So in total and complete frustration after two seasons of complete failure, he went to an old friend or acquaintance, Dr. Bassett McGuire at the New York Botanical Gardens, and explained to him what he was trying to do. "I'm ready to throw in the towel unless you have a suggestion of what can be done."

McGuire—if my memory serves me right—was in his seventies, and he was head of the tropical studies at New York Botanical Garden. He told Ludwig what he thought he needed to do, and Ludwig said, "Okay, I don't know how to do that. You go do it."

What McGuire essentially told him was that he needed to do a 180 and hire biologists to be responsible for the program, rather than construction people. Having not been there at the time but being familiar with the tropics and being familiar with the Ford effort in Fordlandia, McGuire told him that he needed to hire a pathologist, a geneticist—

Harold K. Steen (HKS): A geneticist. It was you, right?

CEP: That was me. And someone in silviculture and forest management. I think they jointly decided what the person should be and what his background should be.

HKS: Just for the record. You have a Ph.D. in genetics under Bruce Zobel.

CEP: Right.

HKS: Okay. If you're one of Bruce Zobel's geneticists, you would have entree.

CEP: Right. There were three criteria. One was that a person had to be a self-starter and someone that had gone somewhere and started a program on his own.

HKS: You had done these things?

CEP: Yes.

HKS: Give an example of your background. I want to know how you were picked out of the crowd.

CEP: Okay. Well, that's one of the criteria. I went from N.C. State to Auburn, and from Auburn to Oklahoma State. At Oklahoma State I started the genetics program and a graduate program. I set up a tree improvement research station in southeast Oklahoma so that today the trees that Weyerhaeuser is planting, basically all the pine being planted in the region today you see come from that program.

So the first criteria was that someone is a self-starter, because you're going to be in the Amazon, and there's no one to tell you what to do.

HKS: Sure.

CEP: Second, you had to be young, and young because there at the time were no schools. At the time that they started calling me, I had one young son, just about two years old. It would be at least a few years before there would be an educational problem with the family in the Amazon.

HKS: So the idea was you'd go down with family, a permanent situation.

CEP: Right. It's permanent. The third thing, which I never agreed with and generally don't, is that the person had to have a Ph.D. to be qualified.

HKS: It's interesting that Ludwig would require that.

CEP: That was McGuire.

HKS: McGuire. Okay.

CEP: Essentially what that did was rule out a lot of qualified people because they didn't have a piece of paper.

HKS: Yes.

CEP: Once I went to New York and they explained the program, explained the objective and what they were trying to do, then I knew that I couldn't walk away.

HKS: It was really attractive to you in a scientific career sense.

CEP: Right. Right. It was a terrible thing to do because I had sold a fairly significant genetics program at Oklahoma State and had many

McIntire-Stennis research projects and had gotten a Ph.D. program going with all the students I could handle. And just to have succeeded in doing that and having built new greenhouse facilities and new research station and everything, to walk in and tell the experiment station director "good-bye" was a terrible thing to do.

HKS: Did he think so?

CEP: He thought it was a terrible thing to do, and I was chastised. [chuckling]

HKS: So you really were burning a bridge in that sense.

CEP: Right. I burned a bridge. I have, and always have had, a burning desire and a concern for man and what happens to man in this world and for the masses of starving people that unfortunately we in this country never see and really don't want to see.

The Jari project is in a totally isolated region of the world, and still, there's no road connection with the outside world. A river trip from a city of any size to the project is two days.

HKS: What kind of a boat? Is it comfortable? Or primitive?

CEP: I would say it is comfortable because you can find a place to hang a hammock.

HKS: All right.

CEP: To most anyone here, it would be horrible. The only other option was air, and of course we had to provide that ourselves. Once they had explained to me what the objectives were, I went to Brazil to look, and once I looked I was really hooked. They had cleared probably thirty thousand acres by the time I arrived.

HKS: There was a superintendent or a manager on the scene, on site.

CEP: Construction.

HKS: Construction. So construction was running the show.

CEP: Construction running the show. When I arrived, construction was still running the show, and I was reporting to a construction superintendent.

HKS: That's Morrison-Knutson.

CEP: Morrison-Knutson was in early, and they were in on the first failure, and they were gone. Morrison-Knutson left, but many of their employees stayed.

HKS: I see.

CEP: I was reporting to a construction superintendent named Boyd MacMillan. As a equipment and construction superintendent, he was an excellent guy. When I arrived there in September of '69, they probably had cleared thirty thousand acres and did not have a nursery. They had a small nursery, in camp, but they did not have a nursery to produce the trees to plant on this thirty thousand acres.

Contrary to popular opinion, when you clear and burn the jungle, it doesn't turn to desert. You clear and burn. You do not kill. I mean, heat rises in a fire, and so you don't kill the roots. And the jungle comes roaring back. Within that region there's a dry season and a wet season. Land is cleared in the dry season and burned at the end of the dry season, so immediately after it's burned, it starts to rain. Because of the fire, you have a release of a tremendous volume of nutrients and plenty of water, and so all this vegetation comes roaring back. Within a few months, you have an absolute green cover that is almost difficult to get through.

HKS: So the plan was to use native cover, native species, commercially? Or was it just clear the land and plant whatever you're going to plant? I mean, obviously you had to clear something to have plantations.

CEP: Well, it's both. Let me back up a minute. When I arrived, they had land cleared, but no trees to plant. And so of what value is a geneticist if there's no plantation? Of what value is a geneticist if you don't have trees in a nursery?

HKS: What's the flaw? McGuire had a plan, right? And Ludwig says, "Make it work." But the plan had some flaws in it, apparently.

CEP: Well, they simply didn't understand what was going on and what the position was. Now, to back up, the next stage after the engineering and construction firms left, he hired a consulting firm. There's an outfit in Vancouver, Florestal, which is an international forestry consulting firm.

HKS: This is B.C.?

CEP: Yes. And they sent a fellow named William Oudshoorn. He, in all likelihood, was an excellent forester. He may have known what to do and how to do it, but with a strong-willed construction superintendent kind of boss and being a consultant rather than employee, what could he do?

So it's easy to criticize Florestal and the people they sent, I saw there were all kinds of things wrong. But in reality they probably did not have a framework under which they had a possibility to succeed. So it failed another year, but in no way should it be laid at their feet.

I rapidly came to the conclusion that I had left an ideal situation in teaching and research, and here I was in the jungle with thousands of unknowns. The only way it was going to function was for me to be able to do what a forester knows should be done. I jokingly say that I became a hero because I knew that you plant trees with roots down instead of up. And I knew that you planted when it rained instead of when it was dry.

So that's the stage it was in. It was so basic. Yet, from the construction side, those basics not understood.

HKS: Almost like being in the Peace Corps.

CEP: Correct. One of my early bosses, after the construction guy, was an engineer named Robert McPhail, who had been president of Kaiser Brazil. There was no plan before, and so he had a plan. As engineers do, he scheduled, and he forced the following of that schedule, whether the environment matched that schedule or not. This led to another disaster.

It became clear to me that the only option I had to survive was to proceed and do what I thought needed to be done biologically, in spite of what anything any of my superiors said.

HKS: Your superiors were there, not in New York someplace?

CEP: I had a superior there who was either a construction superintendent or an engineer.

HKS: What do you think they thought of you coming down there in the first place? It was a good idea?

CEP: It was a good idea, and it was fine. Personally, they liked me. But they couldn't stomach insubordination, which is not a surprise. I guess this is the point to recount some of the things that happened.

Once I saw the situation, I struck out into the jungle and found the best place for a nursery, and that is not easy. One thinks that the Amazon is flat, and it's not. You need a flat place for a nursery. And with the flattest place I could find, we still had to spend a tremendous amount of time and effort in putting in wooden drainage ditches so that the nursery wouldn't wash away.

We had a central drainage system, and we had a wood-lined drainage ditch between every three or four rows of the seedlings. We spent the first few months building that nursery, so we could produce thirty-five to forty million seedlings almost immediately.

HKS: You were able to convince the construction people to give you some equipment.

CEP: Not really. They were all for a nursery and understood that a nursery was required. But they wanted to build it in a construction manner, which I would not agree with. We used the equipment weekends, nights, when I could steal it from road construction or bridge construction or wherever. Let's put it this way; they permitted us to steal the equipment when they weren't using it.

HKS: [laughter]

CEP: [chuckling] But there was no program change as far as land clearing with equipment or road building or anything else in order to allow the construction of a nursery. So we got the nursery built. I believe we sowed the first seed the tenth of October. We had to be planting the first week in January.

Building this nursery included the clearing, the drainage system, irrigation system, because this is the dry season. We had a complete irrigation system. The first planting season I was there came along, and our seedlings were maybe average eighteen inches tall. And the species was *Gmelina arborea*.

I concluded that the seedlings were too small to plant and too succulent to go into the field in that particular year. The rainy season didn't start on schedule. The executive director of Jari at the time was an engineer, and he had scheduled planting on a given day. We had to start. We started, above all my protests. I was not responsible for planting.

HKS: The construction people were planting?

CEP: No, we had a forester who went down the same time I did.

HKS: Who was that?

CEP: Don Cole. He had spent most of his professional years with Continental Can in Georgia.

HKS: So you were the scientist. Then they brought in a practical person to make it work, in that sense.

CEP: Right. Don and I could have a conversation, "Don't do it." But his boss said, "Plant," and so he planted.

I'd go a week behind the planting crews. The seedlings were so young and succulent, and it hadn't rained, that they were all dying. It was going to be another disastrous year, and we were already faced with planting small succulent seedlings in areas that had been cleared on the wrong schedule. Rather than planting these small succulent seedlings on bare ground, they were being planted in vegetation that was roaring back from the original forest.

I knew that it wouldn't work, so I begged and pleaded and fought, and absolutely no one would listen. Of course, all this time, Ludwig in New York is checking every day to see how planting is going, and he's being told, "We planted a thousand acres today and a thousand acres tomorrow, and everything's going fine."

I calculated the nursery was much more successful than I had dreamed, and we had twice as many seedlings as required to cover the land that was prepared to plant. I knew everything they were planting was dying. When the nursery inventory got down to where we still had enough left to go back and replant everything, I went up to the nursery one morning and, rather than lifting and shipping on schedule, I sent all employees home and hid all nursery equipment in the jungle. This forced an end to field planting.

You can imagine the height of insubordination, the wailing and gnashing of teeth. The organizational structure at the time was that Ludwig was in New York. He had a fellow named Francis Thomas who lived in Caracas and was a mining superintendent. Ludwig had mining operations of various kinds scattered over the world, and Francis Thomas was responsible for the mining group. And he also was responsible for Jari. Then, technically speaking, McPhail reported to Thomas. Even though McPhail liked me personally, there was no possibility that he could allow that kind of insubordination and maintain any semblance of organization or authority over the program that he was responsible for.

HKS: It wasn't just you, you might be a bad example for the rest.

CEP: Right. I was told to pack my bags, and took my family and went to Caracas to explain my actions to Francis Thomas before he sent me on back to Oklahoma.

HKS: You didn't have much going for you in Oklahoma.

CEP: Right. [laughing] So I went to Caracas. Fortunately, he wasn't a violent man, but he was verbally violent. I sat there and listened. Francis Thomas at the time was nearing retirement, and here this young whipper snapper, who probably at the time I was thirty-four, thirty-five, and he just, he couldn't contain himself. He fired me I don't know how many times during that session. I guess what stirred him up more was that I didn't get excited. I didn't argue with him. I just listened. When he wound down and was exhausted and had nothing else to say, then I started.

I essentially explained what was happening, and he said it couldn't be. I dared him to wait a month, and I'd go back to Brazil, continue to do what I could do, and if I was wrong, then if there were any seedlings surviving on everything they had planted, then he didn't pay my salary that month.

HKS: So you were again predicting that there would be a total wipe-out.

CEP: I knew it was a wipe-out. They were dead when I left, but nobody would face it. I could take people to the field and show them, but the boss said, "Plant." And so I knew they were dead.

I said, "If I'm wrong, you don't owe me for this month. But if I'm right, then we still have enough seedlings to go back and replant and have a successful year." Well, he wasn't there and he couldn't see, and, with his background, like others, he wouldn't be able to tell whether a seedling was alive or dead.

HKS: Was he reluctant to go on site, or would that be a violation of some protocol?

CEP: No, and he did. I went back, and then, within a week, he came. By then, the first planting was like a month old. You could pull a seedling up, and the bark stayed in the ground, and a bare stick of wood came up because it was rotten. Then it was obvious.

So that was the first time of many that I just simply ignored an order, did organizationally absolutely the wrong thing, made people unhappy. Unfortunately, in those situations, people get hurt because then the forester responsible, it became his fault, even though in writing daily, there was a directive that told him what to do and when to do it.

In the end, he was asked, "Didn't you know it was too dry to plant?" Ludwig asked that question. "Yes, I knew that." "Well, why did you do it?" "Well, my boss told me to." But in a Ludwig organization that is no excuse. One of the quickest and most fatal mistakes that one can make is doing what Ludwig tells you to do if you know it's wrong.

HKS: So that's no excuse for Ludwig.

CEP: It's no excuse. The fact that he told you—

HKS: What happened? The forester got in trouble. How about the engineer who made the plan to begin with?

CEP: He disappeared, too. But not immediately. The forester took the rap. The next most critical thing was to get good, vigorous seedlings to cover the ground, because once that jungle was cleared and burned, if you don't get it covered immediately, you spend an absolute fortune in plantation maintenance. The root system that's already there is well-established, and any seedling you plant is at a disadvantage. So my immediate objective was to get as strong a seedling as we could get. We planted on schedule. The nursery functioned extremely well.

HKS: You got an adequate-sized seedling in six months?

CEP: Less than six months. My boss was Robert McPhail, on site, who was the engineer. I had learned that anything that appeared drastic that I needed to do, I couldn't talk him into it. We had seedlings up about ten feet tall, absolutely beautiful.

I didn't know he was bringing them. It wasn't my position to know. But he brought a group of visitors, and unfortunately, the experience was so traumatic I don't remember who the visitors were. But I had taken a heavy-duty sickle-bar mower and mowed down all the seedlings in the nursery to where they were just stumps, about six inches tall.

Now, if you can imagine, in our country we plant seedlings when it's cold, and the seedling is dormant. Well, there is no dormant season there. In the planting season, when it begins to rain, yes, it rains, but you still have most of the hours of the day with sun, and the transpirational loss is tremendous.

So if you have a seedling ten feet tall, it's cumbersome to handle, but in addition you have a horrendous transpirational loss. Unless it rains continually, the chance of survival is not good because you just don't have enough root system to support that much leaf surface area.

Without permission from anybody, because I knew I wouldn't get it, I mowed down the nursery, and we had just finished carrying off the tops and throwing them in the woods when this caravan of vehicles shows up. And he's explaining to them all along how successful the nursery is, and how beautiful it is [chuckling]—

HKS: [chuckling]

CEP: The seedlings are ten foot tall! [laughing] Comes around the corner, and they're all gone. Again, I was called on the carpet, and the only thing I said was, "Remember last year." We succeeded in replanting and getting better than ninety percent survival on all the land that was prepared to plant.

HKS: Tell me again how tall these seedlings were before you mowed them.

CEP: About ten feet tall.

HKS: Ten feet. So it was kind of spectacular.

CEP: Absolutely spectacular, and beautiful. You have such a good feeling. We went through a few weeks of horrible times, because here I had maliciously destroyed the nursery. But by that time, they were afraid that I was right. They were all ferociously mad at me, but afraid that I was right, and so they didn't put me on an airplane.

Well, planting season came. We lifted the seedlings, and a seedling in that condition is just a stump. You can throw it right on the ground, and leave it in the sun for several days without killing it. So the care required to plant didn't exist. I mean, you could lift all these roots, put them in big boxes, haul them to the field, and put them in gunny sacks, and the guy carries a gunny sack and plants all that.

HKS: Hand-planted.

CEP: Hand-planted. There were two times in our experience there that we had a hundred percent survival. Out of about twenty-two thousand acres planted, we didn't lose a seedling.

HKS: Wow. Not even loblolly pine does that well.

CEP: Right. [chuckling] And because they were never under stress, because the transpirational loss was not a problem, they started growing immediately. We set a limit—we had plenty of seedlings—we had to have at least a half-inch diameter root collar for it to qualify to go to the field. They had plenty of stored food, and so they set root and set sail.

HKS: How did you know these things? Was it just scientific speculation on your part about picking the tops off to reduce transpiration loss, and so forth, or was there literature from a tropical forest that you could use?

CEP: I had no literature.

HKS: You just figured it out.

CEP: I just figured it out. Knowing to plant trees with the roots up instead of down. In hot weather, hot sunny weather, any plant with tremendous leaf surface area is not going to survive in a field of planting. Yes, you could put one in your yard, and it would wilt some, and you could put some shade with it and water it three times a day and get it to live. Which they had done. But millions? Planted by people that never planted trees before? In circumstances where they're fighting roots and everything else? No way.

HKS: How big was the root? How big a hole to put the seedling in?

CEP: We took a standard tool, and planting mattock the locals used for planting manioc, and just made one lick, pulled it back, and put the root in and shoved dirt back in and stomped it, packed it, and that was it.

Unfortunately, I had to make those kinds of dramatic decisions in order that I, and the project, survived, because Ludwig said that if there was another failure, he was gone. It wasn't going to work, and it wasn't worth it.

HKS: So the investment he had made up to that point, the mill—that was way off in the future.

CEP: Yes.

HKS: You were clearing land, building more facilities, getting ready.

CEP: Right. The objective was to prove the biological viability of the program, and then think about facilities. That may be enough examples. I could go for a day on those kinds of examples.

HKS: Okay.

## Ludwig's Outlook

CEP: Maybe this is the time to go back and give a little background of why this project and where Ludwig was coming from.

HKS: That's good. The literature features him as kind of a bizarre individual. Maybe because he was reclusive.

CEP: He wasn't reclusive. That's a term applied by basically the media that he didn't give any time to. In circles of friends and associates, he was not reclusive at all.

HKS: So when you read "reclusive" in *Harper's* magazine, it means he wouldn't talk to the author.

CEP: That's correct. That's all it means. Ludwig was not reared on a farm, but he had a yearning desire to grow things. A lot of people think Jari was the first thing he did in agriculture-related areas, but it is not. At the time of Jari, he already had a world-scale cattle program in Venezuela, called *Hato La Vargarena.*

And one of his many endeavors was that he figured a way to dredge the Orinoco River, in order that the tremendous deposits of iron ore could be moved to the Lake States, to processing facilities. The Orinoco was too shallow for large vessels, so there was no economic means of getting iron ore to the Lake States. So, being a dreamer, he dreamed a way to economically dredge the Orinoco, which he did. Having dredged it, then his ships carried the ore to the Lake States.

So having been involved in Venezuela, he started this large cattle program. He also had a program in the Chiriquí Province of Panama. The company there was called Citricos do Chiriquí, and it was a large-scale orange plantation, with processing facilities. He also had a fresh melon program in British Honduras. The objective in that program was to airfreight fresh melons to metropolitan New York, Philadelphia, Boston, on a daily basis. So he was not new to things biological.

He really didn't analyze. He had an uncanny ability to see and project what was going to happen in the future. His basic tenet concerning Jari was that there was going to a major world shortage in fiber, and that he could help prepare for that. He predicted the year, and in terms of start-up of our pulpmill, he only missed that projection by a year.

HKS: In an ordinary corporation, by my perception, you have a fleet of accountants saying, "Well, if you're going to dredge your river, then get your boats up, that's going to be "x" million dollars capital investment, carried at a certain percentage, and you have to have so much income stream you know—

CEP: None of that existed. Most of the economic criticism of Ludwig and Jari is not valid. From a normal corporate world standpoint, from a banker's standpoint, yes, you could criticize. But one has to remember that, at the time when Jari was being established, Ludwig probably had more liquidity than any individual in the world. It was his money. He didn't owe interest on it. He didn't have a board of directors. His idea of fun is proving an idea.

The bottom line proof of that idea is that it is economically sound. If you looked at the history of the development of his empire, if that's what we want to call it, he would put his money into the program until it became economically sound, until he proved that it would work. Then he would turn it over to accountants and finance officers and so on, and they'd go the normal banking route and whatever. It would become a normal business, with banking relationships and everything else.

But those ideas and concepts that he had that were non-bankable, I mean, no banker in his right mind would do what he did. No corporation. No corporation, CEO, Weyerhaeuser, anybody else, it doesn't matter how large they are, can take shareholders' funds and make that kind of an effort. He was a dreamer. He was an idea person, and he was a rare one that had the economic capability to proceed on his ideas.

HKS: Might he have agreed that he would have made more money in the long run if he'd bought IBM stock? He obviously made a lot money. Or would he say, "No, I'm really making more money this way."? And he could prove it with his bank account?

CEP: His objective in life was not making money.

HKS: Proving ideas.

CEP: Even though he did. Proving ideas. There will be some things that I will not disclose, but there were many occasions when there were easier ways to have done what we were trying to do. It included buying controlling interests of a major forest products company, which under the given circumstances of the day, the stock values of the day, and so on, he could have done, and never missed the money. In so doing, we would have had the in-house engineering, mill operation capability, everything else.

So from a pure economic standpoint, that would have been the thing to do. He would have made a tremendous amount of money, as it turns out, having done that. That is an example of where just simply making money was not his objective. The bottom line reason that he did not do that was that he was afraid that in so doing it would dilute the focus of his objectives.

HKS: Interesting.

CEP: He felt that there was going to be worldwide shortage of fiber, and so he set out to find a place where he could establish a very large plantation to contribute to these fiber requirements. Being the kind of person he was and the stature he had, he had a lot of friends in the forest products industry. He had friends in many industries.

Probably his closest friend in the forest products industry was Reed Hunt, who was CEO and president of Crown Zellerbach. I won't go ahead and name the others. He had friends to talk to. He wasn't just out scatter-shooting at the world, trying to find a place.

## Selecting Gmelina

CEP: He had an employee, an aide; his name was Everett Wynkoop. He had a young Panamanian agronomist named Juan Ferrer. Juan had helped establish the citrus program in Panama, and had helped establish the melon program in British Honduras, and was a, I'll say, a consultant to the cattle program in Venezuela. Mr. Ludwig sent Wynkoop and Ferrer, essentially turned them loose, to help find a place.

They looked at Nigeria, but if you remember the time frame of the Biafra War, there had just recently been millions of Ebo people killed. The land was there. Everything was there. But he was concerned, and justifiably so, about the political stability. So he backed out of Nigeria.

One of the reasons he was first looking at Africa was that, over the years, gmelina, the tree that he had chosen, had been planted in many countries in Africa. A government would come in, and they're going to do everything right, and they need to replant and have wood for charcoal for people to cook with, and so on. They would plant, but there was no money for plantation maintenance, and the thing would fail.

HKS: So the basic properties of gmelina were fairly well-known.

CEP: Somewhat. It was originally from Burma, and the slopes of the Himalayas, and it's an extremely fast-growing tree.

HKS: Fast-growing there, too.

CEP: Yes. And has an excellent fiber for fine printing papers. So he chose gmelina, and the search was on to find the place to grow it. When he learned more of what was happening in Africa, and it was getting increasingly difficult to get into Burma, he sent Wynkoop on a seed-collection tour. To see this guy in the office in New York, with his suit and tie and all proper, you could never imagine that this same guy would roam all over Africa in any circumstance and set up seed-collection stations in whatever tribe, at whatever village, anywhere on the face on the earth. It was rather startling.

HKS: Did he keep track of provenance?

CEP: Yes.

HKS: So you knew where he got the seed, elevation and all the rest.

CEP: The problem was that all the seeds from Africa came from the base of the Himalayas somewhere. So you didn't know. You knew it was in Africa, but you didn't know where it really came from originally.

Being fearful that it might be long-term in finding a place for his large-scale planting, with Burma closing and Africa in turmoil, he sent Wynkoop on a seed-collection binge. They collected. Ludwig bought a small area in Costa Rica, adjoining United Fruit properties, and put in a planting, using seed from all the different countries in Africa, in fact seeds from anywhere they could get it, so that he would be guaranteed of a seed source when he did find the proper place.

HKS: Do you think this was Ludwig's idea or McGuire's advice? I mean, that he would have thought all this through.

CEP: This is probably Ludwig. The details were probably Juan Ferrer's. Ludwig had the capability of thinking up all of the "what if's" and planning for contingencies. Once he had an idea, failure was not in his vocabulary. Once he embarked on this, he then could think of all contingencies and what needed to be done. In all likelihood he's the one that said, "Hey, with everything happening in the world politically, we have to secure a source of seed, and so go gather it."

HKS: He had the resources to do it.

CEP: Right. And he had the economic and political connections to be able to call the president of Nigeria and say, "Hey, I've got this problem, and I'm going to send this guy. Let him go wherever he wants to go." And he could go wherever he wanted to go.

HKS: He had that because he was shipping a lot of oil around the world.

CEP: It wasn't only oil. It was oil, mineral.

## A Site in Brazil

CEP: At the same time that they had given up on Africa, they were collecting seed, but he had started looking in Brazil. Now, politically, the last revolution in Brazil was in '64, and the new president after that revolution was Castello Branco. And, of course, in our terminology it was a military dictatorship and not a democracy, and so it's bad.

Brazil has a very similar form of government. They have a president. They have a cabinet. They have a House of Representatives. They have a Senate. They have a Supreme Court. It's essentially modeled after us. At the time, because of the revolution, the military dictator is superimposed over the normal democratic system, so that when the democratic system gets out of the framework of what the dictatorship is willing to allow, then the dictator says, "No, you can't do that."

Even though it was a military dictatorship, House, Senate, Judicial, all these things functioned. Castello Branco was president, and he had a what we would call a secretary of planning and development. There it's called a minister level. His name was Roberto Campos.

They needed economic development, and Ludwig started talking to them about a location in Brazil. He needed a place in the tropics—they were interested in developing the Amazon.

HKS: Gmelina is up in the Himalayas but it's still tropical.

CEP: It's tropical.

HKS: Humid tropic.

CEP: Right. And there was really no limit of size in the Amazon, so they essentially gave him a green light that any property that he could find, he had tacit approval of the government to proceed.

HKS: Was the property owned in the Amazon, was it owned by the state or privately owned?

CEP: It's private. To put it in perspective, there were four governments in succession that we had tacit understanding with all of them that Ludwig would put in the seedlings and prove the economic viability, building schools, hospitals, roads, airports, all of the infrastructure.

Ludwig realized from day one that no company, whether it's oil, gold, whatever, no company long-term, forever, can be the mother, the father, the doctor, the lawyer, the priest to everybody all the time. Economically it does not work. And so the understanding was that he would put in the front money to make it work, build the infrastructure. Once it was functioning and it was economically sound and we knew for sure that the jobs were there forever, then the government would come in, and we would give them the school system, the hospital, the airport. At that stage, rather than us paying a hundred percent of the cost of the hospital, the hospital comes under the normal Social Security program, the schools come under the normal program, the company would pay property taxes and taxes on inventory, and everything, all taxes that a company normally pays, which would pay for the school system and medical services and everything else.

That was the basic key and the basic understanding of he being in Brazil. Very few people know it, but that is the reason that he left.

HKS: Various articles mentioned tax relief. This would make a lot of sense. I mean, he's putting infrastructure in, so why have him pay taxes?

CEP: Right.

HKS: It looks different the way you're saying than the way the articles portrayed it.

CEP: Right.

HKS: Like they were trying to lure him in with tax write-offs. But it was—

CEP: If you have no profit, there's no taxes.

HKS: Yes.

CEP: We had a tremendous amount of incentives. There were incentive programs for forestry, incentive programs for industrial development, all kinds of incentive programs that we utilized.

HKS: They were available to any investor in Brazil.

CEP: To any, and we used those. But to go back to our beginning story, so once Ludwig had had these meetings with Castello Branco and Roberto

Campos, and because he had been shut out politically in other areas that he had picked, he decided that he liked the Amazon.

The land ownership in the Amazon—a lot of the titles go back to crown lands. The language in Brazil is Portuguese. Brazil was controlled by Portuguese. The original land grants in Brazil are crown land grants from the king or queen of Portugal. So Portuguese families who came to Brazil, and in this case came to the Amazon, received crown land grants from the king, from the throne in Portugal.

HKS: From that time on, it's been private ownership.

CEP: I've studied and worked through all of the land titles, and like would happen in this country when there was a family that got crown lands, the neighbor couldn't hack it, and he had crown lands, and he went back to wherever, and he sold what he had to his neighbor. There was one family stayed and put together this massive piece of property from other owners over a period of more than a hundred years.

HKS: Did they develop any of the land?

CEP: In terms of the economy of the day, yes. They built, the literal translation would be trading stations, up all of the tributaries. The main two rivers that they controlled were the Jari River and the Parú River to the west. They built trading stations up these two rivers and the tributaries, and so it was a collecting economy.

HKS: What did they collect? Medicines?

CEP: Their number one economy was Brazil nuts. The region has a tremendous Brazil nut tree population. Number two is balatta. Balatta is the latex—there's two forms. The main one is the rubber tree, rubber tapping. The next one is a latex from a tree called massaranduba. They collected seed that some French firm would buy for perfume. They'd collect hides, jaguar teeth, anything that the outside world would buy.

This family in boats—they had a fleet of boats—would bring in cloth, hand tools, ammunition, pots and pans, salt, sugar, basic needs of the people that lived in the region. They traded things that people needed for the items that the people extracted from the forest.

HKS: There was obviously enough volume of business to make it worth their while to buy all that land.

CEP: Right. "Buying" the land is probably a misnomer. They took the land, in most cases, when it was abandoned. It's like today. I can buy for you any number of acres you want in the Amazon for almost nothing. We at times in my present company have owned large acreages in the Amazon because it was expedient from the standpoint of the banking system in Brazil. But never show it on a financial statement.

HKS: But there's an owner that you buy it from.

CEP: You agree to accept title so that guy isn't hung with taxes that he's not paying anyway. But as a non-Brazilian entity, you have to pay taxes.

HKS: Brazilian sovereignty over the Amazon is rather flimsy.

CEP: Correct.

HKS: Even today.

CEP: Even today.

HKS: There is the state of Pará, and there's the territory east.

CEP: Amapá.

HKS: That would be even less developed.

CEP: Even less.

HKS: Same as our territories in the States.

CEP: That's correct. It would be like going to back to Oklahoma and not seeing a white man. Yes, there are people there. Yes, there are activities there. But there's not enough activity to justify a form of government other than a federal government saying, "Jose, go to the territory up there and try and have some form of law and order." That's about it.

So when we owned land, we didn't control it. I mean, anybody could come in, collect nuts, cut down a tree and haul it somewhere to get money for some medicine for his kids or whatever. There's no reason to control it. The cost of control is much more than the benefit that one would reap from the land.

HKS: When the articles I read referred to "squatters," that's really not an accurate description.

CEP: It isn't until, like Jari, you start developing the asset on a world-class economic base, and then "squatter" is a real term and a real problem.

# Purchase of Jari

CEP: Robin McGlohn was one of the original Pan-American, international pilots. When Pan Am would open a new flight somewhere, Robin was one of the guys that would open it. So the first time Pan Am flew from U.S. to Rio, Robin opened it. He opened many of the new routes in the Pacific. He was the first in the old flying boat.

After the war, he picked up everything he had and moved to the Amazon. He was a well-known folklore legend in the Amazon. Georgia Pacific bought some property and a mill in the Amazon, probably in early '60s, late '50s. They bought it from Robin McGlohn.

HKS: They bought it from him.

CEP: Right. He put the property together, he built the mill, and he sold it to GP. He was continually doing that kind of thing. Robin McGlohn did the intermediary work of putting Ludwig together with the Andrade family that owned the Jari property. In one way you could say Ludwig bought the property from McGlohn. He bought it from and through McGlohn.

The old patriarch in the family that owned the property, like happens in so many families, they had fought and struggled for a century, and they had Brazil nut processing facilities, where they exported Brazil nuts all over the world. They had heart of palm processing facilities; they shipped all over the world. They shipped other extracted items throughout the world. So it was, at that day and time, a fairly substantial family. But they had enough money that the younger generations were ruined. You know the old saying, "a shirt sleeve to shirt sleeve."

HKS: Right.

CEP: There was really nobody capable or wanted or cared to take over the helm. So the old man sold it. The first offices that Jari had in Belém were the offices of the parent company. When Jari was purchased—and Jari is not a new name—when Jari was purchased, he bought the boats, the land, the stores out in the jungle, the whole deal.

If you, in a discussion, told him that he did things because of his concern for other people, I mean he was a rough, tough, knock down drag out kind of a guy. If you told him that, he'd deny it. But throughout the years, time and time and time again, I saw him do things that said, "I'm doing this because of my concern for people."

When he bought the property, the logical thing for him to have done was stop all the trading, close down all those stores, and gone about his activity. But there were too many people involved. For probably eight to nine years, we did slowly slow down, stop all those extractive activities which over time were picked up by the competitors of the original Jari. But he did not just simply walk in and drop everything and leave those thousands of extractive type people destitute with no market and nothing to do, when in reality it was of no economic advantage to him whatsoever.

HKS: Was there a labor market there that some of those extractive people could have been hired to work?

CEP: We created the market.

HKS: For those who wanted to settle down and get a job, there was a job there.

CEP: Right. Over time, we hired many of them. But those who chose to remain in that form of life went to the competitors of the original Jari company.

HKS: The fact there were these relatively minor things going on there wasn't a part of Ludwig's plan. He inherited those.

CEP: That's right.

HKS: The write-ups. Some of them suggest diversification, that Ludwig never missed a bet to make a buck, because he had Brazil nuts and so on and so forth. But he wasn't really interested in that.

CEP: That's correct. A sidelight to that. We had, over the years, developmental programs in various things that would be additional agricultural and industrial activity for Jari. He had friends in New York, in IFF, which is the International Flavors and Fragrances, which is a very large company.

We worked on miracle berry and patchouli and all kinds of activities like that that were a possibility of development on an extremely large basis, supplying companies like IFF. Those extractive activities that were there that we continued for years were not continued from a profit motive because there was none. They were continued to allow the people time for us to develop and them to either go to work for us or continue in what they're doing and work for other people doing the same thing.

## Ludwig's Age and Liquidity,
## His Manner and Management Style

HKS: How important was his age? He didn't know how long he was going to live, but clearly it was a time of life where he didn't have a lot of time left. Others have said he was in a hurry.

CEP: It was extremely important. The first day I met him, he said, "I only have ten years to live, and we have to get this done." Every year throughout my experience with him, he only had ten years. So after the fifth year, he had ten years. After the tenth year, he had ten years.

HKS: [laughing]

CEP: Up until he became an invalid, he probably still had ten years. But he was in a hurry, and he made decisions based upon time. In many instances, he would want us to do something, and we would say, "Look, we can do it, and it will cost a million dollars, and it will take us six months to do it. Now, if you'll let us do that over a three-year span, where we can learn a little bit, step by step, as we go, it will be cheaper, and the end result will probably be better." He'd say, "Forget it. Do it in six months."

Many, many, many occasions, those kinds of discussions occurred and, without fail, that was the kind of decision he made. Yes, he was in a hurry. You also have to remember that he had tremendous liquidity that was growing by leaps and bounds every day, and he needed to do something with it. He was a penny-pincher in almost all respects. He didn't throw money away. But applying whatever money is required at an objective, saving time, to him was not throwing money away. He would do that, but if he caught you throwing—you know, half-using or not using the back of a sheet of paper, you were in trouble. I mean, it's waste. So you stop and you pick up a penny. You use the back of a sheet of paper. You don't put half a load in a washing machine. But you spend an extra three million dollars to save two-and-a-half years.

HKS: He obviously had confidence in his basic ability to grasp a problem. Was he vain? Was it ego.

CEP: No. No. Not at all.

HKS: Was he hard to convince if he was wrong? If you thought he was wrong, was he open to suggestions?

CEP: To many people, no. And unfortunately—I mean, he didn't ask. In many instances, he'd come in and he'd say, "Look, fellas, this is what we're going to do."

HKS: I see.

CEP: It could be off the wall and just as wrong as anything you can imagine.

HKS: Did he have a forceful personality?

CEP: Absolutely. Forceful, overriding, demanding. So forceful I have seen some of the world's greatest and strongest corporate executives melt in his presence and eat out of his hand, so to speak. Because of his power. They, yes, ran a large company and they were president and CEO and were revered and respected and everything else, but they didn't own it. Here's this guy, a larger organization than theirs, that has total power.

Anyway, the only way for anyone to survive with him was to discuss and if necessary argue. Whereas many people were fearful of confronting and arguing, I never was. Because I saw too many people commit suicide by simply doing what he said to do. Many times, he'd tell me what to do and in later years, as I learned more, he still would tell me what to do. I wouldn't do it. I would go find someone that could do it, or find the proper person in the organization that it was really his responsibility anyway, and tell him what I had been told to do, and then he would do his job.

HKS: There's been a lot of writing about his managing style, with thirty directors in fourteen years. He was so overpowering, it took a very strong manager. You have to explain why you survived.

CEP: I'd have to go back to the calendar—but I survived more years than all the other thirty-three combined. It was because I'm not smart enough to have fear. I and many others dearly love the man.

HKS: Why was that?

CEP: Respected him. Didn't agree. I didn't agree with his personality at all. One thing that literally slayed many people is that in the proper circumstance, not in a bank meeting, not with the Queen of England, but in a proper circumstance to him, he had the language of a sailor. And when things were bad, his language was absolutely terrible, and he could walk up and just paint the air blue.

Now, to me, he was not railing at me as an individual. He was not calling me all those names. It was the circumstance. If you ever let the thought enter your mind that he was attacking you personally—I mean, there stands my boss and he's calling me all those names and he's blaming me for all these things happening. You'd be gone in a week. You'd collapse. Just couldn't handle it.

He was an idea man, and from an outward perspective it appeared that most of his ideas worked. But he didn't have one out of a hundred ideas that were worth a plug nickel. He was smart enough that the ones that didn't work he could stop. If he spent a million dollars on it and it was a bad idea, there was no board of directors to tell. There was no stockholders to say, "Well, I had this idea and I didn't get your permission, and the company lost a million bucks. I'm sorry." The million dollars just disappeared.

HKS: When you owned your own company the way he did, it was easy to start an idea, but it was also easy to stop.

CEP: It was easy to stop. I have seen, been in on, and I know there were others larger, but the largest that I was involved in was twenty million dollars. He started down this path, and it was in a country that the country required that fifty-one percent of a venture be owned by nationals.

It wasn't their money. He brought them in. He thought they were good guys. He brought them in because there had to be fifty-one percent, and he thought even though they would own fifty-one percent it was still a good deal. They kept trying to run the show and buck what he wanted to do. He walked in and said, "Look, it's my money. This is the way it's going to be. This is the way we're going to proceed. This is the way it's going to be built, and this is your responsibility, and that's what you're going to do." They bucked. And he said, "Then this program is over. 'Bye." And he headed out the door. They said, "Hey, you can't do this." He went to the airport.

One of the reasons he was so successful is that he didn't brood over that loss. Yesterday didn't exist. The word "what if" was not in his vocabulary. As long as I made a mistake, and I made plenty, as long as I learned from it and didn't repeat that mistake, there was nothing so to speak written down in his book, in his black book, against Posey. Now, if I didn't learn, then disaster. But you could make about any mistake first time and no consequences.

Whether it was personal or financial, there was no such thing as "what if." I talked to him about it one time, and he said, "'What if' is one of the greatest sins of man and one of the things that prevents man to recover when he falls and pick up and run again." Say a guy goes bankrupt or he gets fired. He spends all his time brooding over what if I hadn't done this? What if I had done this? What if I hadn't told so-and-so such-and-such? So your productivity and your capability is drained off in "what if" in the past, when the only significance is past is, "Did I learn from it?" He had that ability so complete that he was always looking at tomorrow, nothing of yesterday. The only family he had was a wife. A very strong-willed, red-headed wife who understood him, and he could get beat up and wounded and run over every day and drag home at midnight, just barely get to the door, and by the next morning she'd have him put right back together again. That is one of the explanations and reasons for his survival.

HKS: Two more questions on Ludwig, the man. Did you call him "Mr. Ludwig" or did you call him "D. K.?" I mean, how personal was your relationship?

CEP: Many people called him "D. K."

HKS: I've seen that.

CEP: And thought that meant that they were close to him. I never did. I respected and revered him. If I were calling him to get his attention, I called him "Mr. Ludwig." But other than that, in day-to-day conversations, I really didn't use his name.

HKS: Was he fairly generous with compensation? Did he give you bonuses or raise your salary? How did he view this?

CEP: He was a shipping man—it doesn't mean it was good in all respects—but he had a ship captain's mentality. You hire a captain for a ship. You put him on the ship. The ship leaves. Either the guy can navigate, can do maintenance on the ship, can take care of the crew. He can run it, or he can't. You're not there to make any suggestions. You can't tell the guy to turn left, right, slow down. You're not there. You have to have complete trust in the captain you hire. The minute that you determine that the captain can't do what you wanted him to do, you have to get rid of him. It's too big a risk—as Exxon has learned—it's too big a risk to have a captain that you do not have complete trust in.

So he applied that throughout everything he did. In many circumstances, it didn't apply. But that's the way he functioned. So, yes, when you became a ship's captain you had complete authority and you were well-compensated.

HKS: Except you wouldn't have stock options and the other things other companies would have.

CEP: But you didn't need it because the compensation was sufficient. I say you had complete authority. But in reality you had him second-guessing. You had a steady flow of consultants in every subject you can imagine on your back and second-guessing and sending reports. And you had the normal corporate individuals, in a corporate structure, not knowing what you're doing other than they're sending money down a rathole and don't like it, and they're clamoring to determine what's happening accounting-wise, fiscal controls, everything else. In that respect, Jari was a ship with a captain, and it was difficult for other people to get their hands on it because he kept them at bay.

HKS: And you had some pressures from the Brazilian government scientists.

CEP: Constantly. We had a steady flow of scientific, educational, military, political. In Brazil you have military colleges. You have the joint chiefs of staff. You have the military war college. Well, just like here. You have a group head of the air force, a group head of the navy, a group head of the marines. You have a school for each of those. At the height of our development we had a group of people that essentially did nothing but receive and tour and explain our program to those kinds of groups. It was a significant portion of our budget.

Back to land. After the meeting with Castello Branco and Roberto Campos, he sent Wynkoop and Ferrer into the region to determine its qualifications for what he wanted to do. Ferrer was scientifically well-grounded and knew what Ludwig wanted to do, knew Ludwig's long-term objective. Ferrer's immediate concern was whether there were enough soils in the region suitable for Ludwig's objective.

He and Wynkoop went in and spent I don't remember how much time but probably two-and-a-half to three months in the field, running lines, soil survey lines, taking soil samples and lugging all these samples out to send them to a lab to try and get a feel. Of course they were doing it on a broad scale. I mean, if they were in a flat, they'd get a soil sample. They'd walk a mile and they'd be on a ridge top and they'd get a sample. They spent all this time and got out.

Juan got back to Belém. He called Ludwig in New York and said, "Okay, we completed the soil sampling. As an individual and what I know about soils, it looks great. But it will be another month before we get results of all the soils tests." And Ludwig said, "Well, they better turn out right because I already bought it."

HKS/CEP: [mutual laughter]

CEP: So here these two guys just spent all this time in the jungle, and it didn't mean anything because he had already bought it.

HKS: The information was ultimately useful, wasn't it?

CEP: Yes, because it allowed other people to look at the results and [still chuckling] determine the areas to begin the clearing program. But those were the kinds of decisions and kinds of things that Ludwig's time frame forced upon you, simply because he wouldn't wait for normal, scientific, bottom-line results.

## Brazilian Concerns

HKS: The land purchase itself wasn't a major investment, as I remember the numbers.

CEP: No. No. In terms of the total money applied, it was insignificant.

HKS: When the property was sold in '82 or '83, a million acres or so was in land title search. Was this a real issue? You explained that the land grant titles were kind of obscure, but did anybody ever worry about that?

CEP: Well, yes. There are, and there always have been, a lot of nationalists in Brazil, the fear of somebody taking over the Amazon. If you can imagine something of the magnitude of Jari, if we went over in the middle of Georgia and struck off a few million acres and put a fence around it. We didn't build a fence around it, but it had a psychological fence in that there are no roads. You can't drive around it. There's no way to get there except through our gates, so to speak. You could fly over it, but what could you see? You're employing tens of thousands of people, and clearing land and planting little trees when you already had trees. So if you have a big tree, why would you cut it and not use it and plant a little one?

So there had to be some other motive. Maybe we had found tremendous reserves of gold. Occasionally they were out there with troops trying to find something. Somebody started the rumor that we were taking big Brazil nut logs, and we had this big auger that we hollowed out the log, filled it with gold, and then put a cap on it and were shipping gold out in logs.

HKS: Why would you have hid it? I mean, why were the rumors that ludicrous? Was there a royalty on gold or something?

CEP: Well, no, but just gold is something interesting. I mean, everybody likes to talk about gold.

HKS: Okay. I wondered why you would have bothered to hide it.

CEP: Because it was such a tremendous find, we were trying to get it out and, being foreigners, were trying to get out as much of Brazil as we could. Of course, they couldn't understand how one individual could conceivably have that much money. And if he had it, why would he spend it that way? Why wouldn't he leave it bundled up in a room where he could see it?

HKS: Was there I'll call it a public relations program to cope with this initially? You've said you had a group that did nothing but conduct tours.

CEP: We didn't go out and advertise. The only way we coped with it, we just simply had an open door. Any scientific group, any military group.

HKS: But you didn't invite journalists in and see it and write articles and that sort of thing.

CEP: No. No. We didn't invite them in. We kept them out.

HKS: Was the original land purchase the basic purchase, or did he keep acquiring more in bits and pieces?

CEP: No. One purchase. But see, very few of the land titles were really any good. I remember one title in particular that said, "Start at the confluence of 'x' creek and 'x' river, and go so many leagues, so many degrees northwest." Well, if you did, it put you out in the Pacific Ocean off of Colombia.

HKS: It goes back to the original land grant. They didn't know where it was.

CEP: They didn't know where anything was. Let's take Alligator Creek. On this drainage system, the people that lived there named this little creek, Alligator Creek. There's a drainage system over here, and there's alligators over there, too. And the natives named that Alligator Creek.

HKS: Was this an impediment to investment, or Ludwig doesn't worry about that? He had his couple of million acres, whatever it was, and he was satisfied with that. How many acres were there?

CEP: Well, it depends. There's all forms of titles in Brazil. If you say that he owned the land for all forms of ownership that existed, it was about seven million acres. But some of the land we ourselves never considered we had clear title to. What we, in our mind, had clear title to was less than three million.

HKS: So you tended to work with that three million acres.

## Land Clearing and Saving the Soil

CEP: We functioned within that three million. Brazilian law is such that if, let's say, there's a piece of land within our ownership. We have papers saying we own it, but we really didn't believe it because of the form of ownership. But we needed to develop it. We'd go ahead and develop it. Build a road through it, plant it, whatever. Brazilian law states that if land is developed erroneously or when titles are in dispute, if the title dispute is settled by the court, whoever owns that land, if it wasn't the one that develops it, has to pay the one that developed it, its value, including developmental cost. Which means that there would be no one that would have the capability of paying us back what we spent on it, on roads, clearing, plantation establishment, and everything else.

To my knowledge, there's never been a question on any land that we developed that we had any inkling that it wasn't under proper title, because there was just no one with the capability to pay. So that wasn't a significant risk.

He purchased Jari in '67 and immediately went out and bought from Caterpillar U.S. seven million dollars' worth of land-clearing equipment and road-building equipment. Sent it down and hired construction people and said, "Build this forest."

HKS: So he bought the equipment rather than having the contractor bring his own equipment.

CEP: Yes, because it was permanent.

HKS: Okay. He needed the equipment when they were through.

CEP: Right, but the land-clearing equipment was a mistake, at least the way they did it. That was one of the other things that I did the second year I was there. I parked all of the land-clearing equipment, and this was a tremendous blow to him because he had spent the money on all this equipment.

HKS: More on equipment than on the land itself.

CEP: That's correct. So that was a traumatic experience. We then started clearing by hand to where we didn't damage the soil, and used the equipment for road construction.

HKS: How about stumps?

CEP: Didn't touch them. What we knew at the time was that they had taken this massive equipment and cleared, and in order to have something that looked nice, they windrowed. The key to growth of any plant in these tropical soils is organic matter. The biological activity in these soils is much more important than nutrient levels. People key in on nutrient levels after a burn, but it's really biological activity, and it doesn't matter what nutrients you have. If you don't have a high level of biological activity, then those nutrients are not in a form that's available to the plants. So when they piled and burned, made windrows, they scraped the soil, and they planted trees in the nice clean area that they scraped. All the organic material was in the windrows. Terrible results.

HKS: Could there have been some modification in that process, a different kind of blade on the tractors?

CEP: Right. That's what they were doing when I got there. And the next season, they cleared with equipment, but just knocked down and let it lay. But still a stump sticking up is a lot less impediment to planting and plantation activity than a stump on its side. If you have a tree knocked over, the roots and everything are sticking up, and you have a lot more volume than if you cut a tree off and you have a stump.

When I looked at the economics of land-clearing by equipment in an isolated region, looking at equipment maintenance and fuel costs and everything else, we started clearing by hand. We rapidly developed to a point that we could clear thirty to forty thousand acres in five months by hand, and not disturb the soil in any way.

We had knock-down and drag-outs until I showed how much cheaper hand-clearing was, and it was about forty percent of the cost of machine-clearing.

HKS: Plus you save the soil.

CEP: Plus you save the soil.

HKS: Interesting.

CEP: To finish the initial subject I started on, Ludwig was not new in agricultural activity. This was by far the largest of any agricultural or forestry activity he started. He didn't know what to do and procedures to use, but he was willing to listen to people that knew or at least knew enough and was willing enough to stand up and fight for what they thought would work.

The initial burning was all a disaster. They'd clear, but they couldn't succeed in getting it burned. Once we showed how to clear, burn, and that we could plant and get good survival, then management-wise we really had no competition. From that point on, we essentially had a free hand on forestry development.

We had no limit on applied research, so we had a sizable research staff in genetics, soils, silviculture, management, administration. In fact, we had as many people in applied research to direct us where to go tomorrow than we had operational management. If you imagine, planting gmelina is like planting tomatoes. If you let the weeds grow in your tomatoes—and I can show you where I live—there are bad results, and it doesn't take five years to find out. It's then.

Gmelina grows so fast that an incorrect step, which nobody knew, but an incorrect step in silviculture, you find out. If you looked. And so we looked constantly. A year never went by that we used the same procedures at the end of a year that we were using at the beginning of a year. There was no policy. There was nothing hidebound. We learned as fast as we could learn, and we applied it.

## Tropical Forest Science

HKS: The tropics is, of course, a huge place. There's tropics in Africa, in Indonesia. Could you make use of other people's research that professors and scientists were studying worldwide on soils and so forth. Did you read the literature and find it useful? Or are issues actually so site-specific that you have to do it yourself?

CEP: I hate to say this, and the scientific community would have me shot if they could find me, but unfortunately many people in the scientific community do not look at the bottom line of the question. The scientific community is influenced by fear of the media, and particularly now, fear of the environmental people and the media associated with it. So they get blinded, and number one, they don't ask the right question, and number two, if they ask the right question, they don't pursue that question to the baseline.

HKS: What I was thinking about when I asked that question, the Forest Service has a facility in Puerto Rico, has been there since '39. They have one of the largest tropical forestry libraries in the world. Could you go there and look at that literature and find things that would be useful to you at Jari?

CEP: No. It's a terrible thing to say. It's hard for people to grasp, because in this country we don't live in that kind of world. But we had no limit. Once our nursery had operated a year, and I picked a young Brazilian technical guy to run the nursery, he and I got on an airplane, we made a nursery tour of the U.S. We visited Weyerhaeuser nurseries, we visited state nurseries, we visited tree improvement programs. I gave him a whirlwind tour you can't imagine, so that he would see the objective that he was shooting for.

I didn't tell anybody I was going to do that. I didn't get permission for his ticket. With Ludwig, let's say he was coming in two weeks. The way to get shot, so to speak, is be needing to do something but well, he's coming in two weeks. I'll wait and ask him. If you ever told him that you waited two weeks to make that decision till he got there, you've got a notch in your black book or on the handle of the pistol or whatever. Because not making a decision when you're supposed to know is wrong. Losing time is wrong, and costs money. So normal operating things like that that needed to be done, you just go do it.

We looked at plantation development in Venezuela. I went to Africa. I looked at everything I could find in the literature that might provide something that we could look at. So yes, we looked. We tried to learn, and we applied it. But the question you ask, I still unfortunately have to say no.

HKS: I see on your resume you published twenty papers, twenty articles. Was that before Jari?

CEP: It was all before.

HKS: So you didn't publish—

CEP: Not one since. Unfortunately it's because if you tell the truth about Jari, no one will believe you. If you say what they want you to say, I could publish something every month. I refuse to write what people want to hear, rather than the reality.

HKS: I want to get back to this major commitment to local research. Did you have a lab so you could analyze your own soils on site. Or did you have to send things out?

CEP: One of the first things we did when I got there was set up soils lab. We were running with all kinds of fears. I mean, the world says and the literature says that monoculture is bad. It says that if you clear tropical soils, they deteriorate and the growth goes down. All kinds of wives' tales.

In the scientific community you could go to—well, I won't name places. You can go and talk to tropical experts in forestry, and common knowledge is that the reason people, in shifting agriculture, clear a spot and grow a crop for two or three years and then move is because the soil deteriorates and is no good. And in some people's mind, it is destroyed forever.

HKS: You read those sort of things.

CEP: Yes, and the scientific community believes it. This is one of those areas where they haven't asked the base question. That is as erroneous as any concept that you can imagine. In reality, what happens is that a guy clears his three acres, and he doesn't have a local store where he can go buy some silvicides. If the store was there, he wouldn't have the money for them. So this jungle, this three acres that he cut and burned, comes roaring back. It comes roaring up from the bottom, and it comes in from the sides. Purely and simply, within three years, in his time and effort, it takes less time to chop down another three acres and abandon the brush that he can no longer beat down with his machete and start a new patch. Because the fire does get hot enough to give him some relief from using his machete for a period.

It has absolutely nothing to do with soil fertility. Now, if clearing ruined it, why do you not see, when you fly over, all these tremendous areas that have been cleared in the past? They're not there. They disappeared. One of our activities today is to go back to areas that were cleared in the past and recover products from those areas.

But you can take a guy who is sitting on this spot, where he's growing something to eat, and you say, "Where were you born?" "Well, back down here?" "When were you born?" "Thirty years ago." And so you can tag the age classes of each of these parcels that he cleared. "Where was your father born?" "Well, he was born down here." "Where was your grandfather born?" "Well, he was born down there."

So we have gathered products resulting from the fire where his grandfather cleared a place fifty, sixty, seventy years ago, in products that wood doesn't rot. And you know what? We collect those products in a forest that the normal scientists going by could not detect that it had ever been cleared.

HKS: This might be the time to bring up this June '93 article in *Science*, which I don't know if you've seen or not, but it's on tropical deforestation using Landsat imagery. So what does that mean? They're showing clearings and loss of habitat. They're saying that Landsat was too imprecise. There's less clearings than they used to think, but the wildlife habitat loss is worse than we used to think. But they're assuming that the land cleared is done for.

CEP: Oh, yes.

HKS: This has been peer reviewed, *Science*, a very reputable, not a radical magazine, I would say.

CEP: What happens is that any acre that was ever cleared or ever touched by man is destroyed. And it's almost to the extent like when you, in the summertime, go home and watch the ten o'clock news and there's a fire in Idaho. The way the media presents it, the land is lost, like ten thousand acres slipped off into the sea and it will never be forest again, and it's destroyed and gone.

I'm not saying that land has not been cleared. Some of the biggest mistakes in the Amazon were projects financed by the World Bank and projects essentially financed by the tax incentive program. Swift-Brazil, Volkswagen-Brazil. Several very large programs. But those areas—I mean, it's real. It was cleared. It was burned. It was put to pasture. And it failed.

To me, some of the greatest forestry opportunities existing in the world today are those areas. Because clearing and burning and planting grass didn't kill the original forest. It's back. Not in the same species distribution, but it's back. Those areas have airports, housing, all the infrastructure.

If I were a Ludwig today, I would go into one of those areas and, having the infrastructure in place, start managing. I would choose probably the best twenty species that at least have some chance of use in our world today, based on our wood technology today, and favor those twenty species and accompany the development of this new forest as an economic base, rather than wail and moan and decry the reality.

I mean, it's real. It was cleared. It should not have been cleared in the way it was. It should not have been managed the way it was. But it is forest today. It will be forest tomorrow. Why not, instead of wailing, channel all of our activities to turning that into an economic base that provides jobs? Bottom line is that jobs prevent clearing of native forest.

At Jari, yes we cleared and burned and created jobs, and the number of acres that Jari has saved from being cleared is already several multiples more than we cleared. And it should last forever.

HKS: So the cost of sustainable development is viable, very viable.

CEP: Absolutely. Absolutely. I haven't read this [referring to the magazine], and I don't know what percent of the total Amazon that they're saying is cleared, but—let's see.

HKS: It's in the southern part. The brown was never forest land. Some kind of savanna, as I understand it.

CEP: Right.

HKS: But Jari is where? Here? I can't quite figure out where we are in this.

CEP: I don't understand their subdivisions, because the mouth of the Amazon should be right in here, and there is no river this shape. See, this is another tremendous misnomer. There are literally millions of acres in the Amazon that are natural savannas.

HKS: This is sort of "before and after." This is ten years apart.

CEP: Yes. But these lands were never forest. Well, they were. But in terms of—

HKS: It says, "clouds." I guess it was cloudy those days.

CEP: But in terms of history of Brazil, if we take Brazil since it was just Indians, all the stretch through here is open savanna, grasslands savanna. There are tremendous areas that are savanna. There are tremendous areas along the river systems that are *várzeas* and not forest. Yet all these acres get counted as cleared.

HKS: In ten years this is more cleared land than before. That's the way I read the map, without knowing the terrain.

CEP: Right. But here again, that is not the Amazon. This is not the Amazon.

HKS: I guess this area drains into the Amazon River.

CEP: No. You'd think it is, but it's not. I won't name the institution, but an institution sent me a bulletin to review. The title of it was I don't remember exactly but something like "Useful Species of the Tropics." They were listed in there, and their characteristics and uses and so on, in alphabetical order by their scientific name. First one on the list: *Araucaria angustifolia*. It's not even a tropic species!

HKS: It grows in Chile, doesn't it?

CEP: If it occurs south of the U.S., it must be in the tropics.

HKS: I see.

CEP: None of this land that you see here is in the Amazon. You take this out, take this out. You take the Amazon drainage. What percent is cleared? And then you take the savannas and *várzeas* out, and what percent is cleared?

See, the river, where the city of Belém is, is on the Parú River. It's not the Amazon. People think Belém is on the Amazon. It's not the Amazon drainage. People have an agenda.

HKS: I don't want to get you off this subject, but this is a photograph on page 1905. It's a very straight line. It's almost like the public land survey of the United States. Has Brazil been surveyed? Why are these ownership lines like that? Very straight.

CEP: Land that is being developed, if you're not a squatter, is surveyed. The law applying to land that is granted to settlers, if it's forested land, depending on where it is and so on, a percentage has to remain in forest. That's the reason you see this kind of pattern. Let's say you get a grant for a hundred hectares. Half of it has to stay in forest.

HKS: That's a federal law.

CEP: It's a federal law. Now, you can choose which half stays in. But half of it has to stay in forest, and so that's the reason for this kind of pattern.

HKS: I see. Are there state laws as well?

CEP: Yes.

HKS: Is that a complication for managing?

CEP: Yes, and there are replanting laws. Brazil has excellent forestry laws, but it doesn't have the economic and social and political means to enforce them. We probably need to get into a social discussion before we complete, but this one statement here is basically, land-clearing is a social problem that has no solution outside of jobs and education.

No government has the political ability or will to prevent starving people from walking onto a piece of property that is otherwise not being used, in order for that person to grow something for his kids to eat. I would go further in saying that no scientist, no environmentalist, no anyone in developed countries has the right to doom that guy and his family to starvation.

Things that we have the ability, economically and based upon our industrialized world experience, to help, there is no means of helping until we realize that the Amazon is not ours. The Amazon does not belong to the industrialized world. We cannot apply any law to it, social or otherwise. We cannot impose anything upon it. We have tried to apply economic pressure in terms of the Amazon, and all we have done is made it worse.

I can give specific, horrible examples of where rules that the IMF or banks or whoever applied have created more problems than they solved. So bottom line is it's not ours. That doesn't mean we can't be concerned. It doesn't mean we can't help. But we can't stop what's happening by being arrogant and assuming it's ours and trying to force the starvation of millions of people. And so far that's the approach we've used.

HKS: What Chico Mendes was trying to do to get the system going, extracting rubber and so forth, and the ranchers didn't like that because they wanted to clear the land and make it sort of an agribusiness. That's Brazilian against Brazilian.

CEP: That's correct. I think what we have to realize is that just like in this country, what people in industrialized regions want to do, seemingly, is preserve the Amazon. That's equivalent to saying, when there were pioneers here, preserve the USA and don't cut any trees.

That's not realistic. The only thing in my mind that's realistic is that as people become better off economically, as they have jobs and as they are educated, they have fewer children. Population pressure is what is forcing clearing. I mean, go to Africa. Population pressure is what is stripping Africa of vegetation.

It's hard for the industrialized world to realize that probably seventy percent of total wood consumption in the world each day is to cook a meal. When you look at the billions of people in India, billions of people in China, the people in Africa, we in our industrialized society are a very small percentage. Most of the wood that's harvested is simply to cook a meal.

To me, the approach has to be to work toward relieving the pressure of masses of humanity, and that's only done through providing jobs and education.

HKS: How about the other questions. The earth needs the Amazon rainforest in order to maintain some atmospheric balance. You have to be sort of a scientist to understand that.

CEP: In my Arkansas language, that's hogwash. Everybody out there is talking about it. I have never found anyone that will answer my question or can stand under my question.

A lot of money is raised by environmental groups by convincing a grandfather that his grandchildren aren't going to have any air to breathe because the Amazon is being destroyed. So what I dearly love to do is be in a group and find a biologist, a high school teacher, even, in the group, and rather than me saying how it is, just start asking questions, and let that person respond.

Because number one, a virgin forest is not a net producer of oxygen. So undisturbed, the total Amazon drainage is not a net producer of oxygen. So you ask the biologist, "Where does oxygen come from?" Well, it is one of the net effects of the photosynthetic process. Photosynthesis occurs most rapidly as a result of growth. Yes, there is photosynthesis in a maintenance mode, but the greatest level of photosynthesis occurs as a result of growth.

So a climax forest has zero net growth. If that were not true, the trees would be as tall as the moon, and you could no longer walk through the forest because it would have grown closed. So there has to be a point at which it stabilizes and there's no net growth. Of course that's what a climax forest is.

If there's no net growth, where does the oxygen come from? Yes, it produces some in the process of maintenance, but a climax forest is in constant process of dying. A tree dies, and young ones take its place. All of the leaves that are dropped every year, and the individuals who die. What happens? They decay. What does the decaying process do? It consumes oxygen. So what little net oxygen was produced in the process of maintaining the forest, then is consumed by decay of leaves and individuals that matured and died. So please someone answer where is this tremendous fountain of oxygen? It doesn't exist.

Now, NASA has done surveys. They grid the world at various altitudes, measuring oxygen level. The Amazon is one of the most oxygen-deficit regions of the world. So if one is really concerned about oxygen supply, we should be doing things in the ocean, in the warm oceans of the world, to increase the productivity of the phytoplankton. Because one of the richest oxygen regions of the world is over the warm waters, and it's oxygen being produced by phytoplankton. When they die, where do they go? They go to the bottom, where they can't consume oxygen in the decay process. So there's a net oxygen surplus.

Now, this is a simplistic, short explanation. But I'm still, after years, looking for someone to explain that away from what I just said. And I've never found him.

HKS: I'd like to pursue two more avenues on this, two more lines of thought. One is on the carbon budget and global warming. If you convert this biomass, you burn this carbon, this causes global warming. I'm not saying it does, but we read that it does.

Another thing I read is that rainfall is affected, global precipitation is affected by changing the hydrologic cycle by converting the forest to another kind of forest or to another kind of vegetation. So if you have a comment on that, we can go on to other subjects.

CEP: I submit that the poor guy in the Amazon clearing three acres to grow something for his family is not the bad guy. Because he is creating a situation where growth is rapid. He is producing oxygen, and he is much less a predator upon the earth than you or I driving our car to work today.

But for clearcutting that has occurred, and take Jari as an example, because of the rapid growth of the plantations, short cycle, rapid growth, Jari is contributing much more oxygen to the planet than it ever dreamed of consuming in the burning process. So yes, it contributed carbon to the atmosphere for "x" limited amount of time, but it has been producing oxygen at a very rapid rate for a much longer period of time, and it will continue to do so, than that period where it contributed carbon.

The other part of your question was sustainability. So far, growth rates at Jari have been greater each successive rotation than the previous. All the literature, everything we were told, all the naysayers that came through. It was just the opposite, and there are specific reasons. The naysayers are going based on what they have seen worldwide in other tropic regions. If you go to many plantations in Africa, it's exactly what happened. If Jari were not managed, it would happen.

HKS: So you think proper management in Africa would solve the problem?

CEP: Management is the key. Understanding the soil that you're dealing with. At Jari there's a range of soil from—we have rich—I'm still saying "we." There are rich, clay soils. There are red clay soils, origin basalt. As fertile a soil as you can find anywhere on the face of the earth, grow anything, good permeability, pH of about 6.7, 6.8, I mean literally an Eden. Literally across the creek from it—I mean, you could throw a rock—and the soil is a sandy soil with a pH of 4.3. Both of those soils can be managed for a forest, but management is totally different. The base key to management of soils in the tropics is maintaining organic. If you can maintain organic, you can maintain the biological community. If you can maintain a biological community, you can grow almost anything.

One other area that we might get into is that there has been—I don't have a tie on the numbers any more—but probably a hundred and fifty to a hundred and seventy-five thousand acres of pine planted in the savannas in the Territory of Amapá. The first of that we, meaning Jari, planted in 1973. What I learned in ecology is that there are natural savannas. I now know there are not. The reason they're savannas, and I mean you have a lush, tropical rainforest, and in many areas it goes to savanna just like a property line like you have in this photo.

Number one, that's suspect. But we started planting savannas. We found that growth, in this case pine, is very good and that in two rotations you can convert a savanna back to a tropical rainforest.

HKS: How about precipitation in the savanna?

CEP: The precipitation has been the same as it is in the rainforest.

HKS: I didn't realize that. I thought it was drier, somehow.

CEP: [chuckling] Oh, it's supposed to be. But what happens in the case of pine is that these have been savannas since Indian days way back when, and they're savannas because of Indians. Because of a hundred inches of rainfall, there has been leaching. You get a sandy, loam soil with nothing but a grass on it, and not a very good grass at that, and a hundred inches of rainfall for a few hundred years, and yes, it leaches.

But when you plant pine—and "leaching" is a terminology like "lost," like five thousands acres burning in Idaho and it disappeared. It leaches, but where did it go? Did it go to China? All the way through? Likely not. So when you plant pine, which has a taproot, guess what? It serves as a pump. So in savanna soils, the reason so far that each generation has been better than the first is that this natural pump is reclaiming those nutrients that leached, that in our mentality meant lost forever, and bringing them back to the surface. When they come to the surface, you get leaf drop and biological activity reestablished. You have a good site.

Now, with that site established with shade, the ground is cool, it's more moist, and animals carry in seed from the adjoining rainforest, and wind blows seed in. The third generation in a pine generation. Once you clearcut the pine and are ready to replant the pine for your next generation, the plantation maintenance cost to fight out the invading rainforest is the same as an acre of cleared and burned rainforest.

So after the third generation, if you clearcut the pine and walked away, you have converted the savannas to a rainforest. That's against everything I ever learned. It's against all the naysayers. So that if someone said, "Let's get rid of savannas," which are basically non-productive. The grass on them in the spring—the grass is beautiful, and the cows still starve to death. So if somebody says it makes economic sense to get rid of the savannas and go to rainforest, it's easy to do. Plant pine.

HKS: With the scientific facilities, research facilities you had there at Jari, you actually documented this nutrient uptake? It wasn't speculation? You observed it happening and you wondered why?

CEP: Right. We didn't know. I mean, we were a bunch of young what I called blood-and-guts scientific pioneers. If I wanted to hire a Ph.D., I hired him. If I found the best soils guy in the world in Timbuktu, I hired him. I had no limitation. So not knowing, and having fear of everything everybody is saying, in our soils group we measured every known variable to man. What the soils scientist community knew that day of things that influenced growth, whether it was physical or chemical, we measured it.

We had soils samples of the organic layer and each horizon, if there was a horizon. In each major clearing, we had soil pits so deep that to get in them we had to dig a stair step procedure where a guy could get to the bottom. So we had soil pits as deep as anyone could conceive that nutrients would leach. In some areas, you had a lateritic hardpan where elements accumulated, so that was fairly well-defined. Others, you didn't. We went deep. Some places in the red clay soils we'd go to bedrock.

We catalogued, tagged, measured everything we possibly could so that we could determine what was happening and, in knowing what was happening, know how to change procedures and management and species. See, people looking from the out in, I mean, everybody is a critic. And every little thing that happened at Jari that we started that, based on sound data, we changed, to anyone on the outside it was a mistake.

I don't view that as a mistake at all. We were dealing in an unknown world. We were dealing under a circumstance where "wait" didn't exist. "Wait until you know" didn't exist.

## Gmelina vs. Pine

CEP: In the case of gmelina, when we started, the first day I hit ground there, I knew that we should be planting pine. I just knew that gmelina, which has a spreading, fibrous root system, would not function well in sand. That was probably my first argument.

HKS: One of the articles I read said that you were forceful in explaining to Ludwig—I don't know if you've been interviewed by journalists or not—but one of them quoted you that you always thought gmelina was wrong.

CEP: Well, I didn't know that gmelina was wrong. I knew that in all likelihood gmelina was very site-specific, and I knew that the likelihood that one species would be the correct choice of a very wide range of soils, regardless of origin—whether it was sedimentary or formed in place, or whatever—the likelihood of that was slim. So, from the very beginning, and if we go back to the first discussion I had with him on the subject, he said, "We have to start with something. I have chosen gmelina, and we will start with gmelina until you prove there's something better. And the minute there is something better, then we'll add it to the program."

That reflects back upon the fact that he wanted to run and was not willing to wait. In a program such as that, you're constantly learning, and Jari would not exist if you waited until you thought you knew everything. Of course, even in forestry and agriculture, where we think we know everything and we look back at what we did ten years ago, we didn't know as much as we thought we did. So with that kind of concept, he said, "We have to start with what we have."

I said, "Well, almost for sure, in the sandy soils we should plant pine." He said, "Well, we are not going to dilute our effort front end. We're going to plant gmelina, and when you're more sure of what you're saying about pine, then fine."

HKS: Gmelina wasn't a brand new species. It had been planted in Africa. What was known about it scientifically, in terms of the kinds of soils it preferred in Africa, or pests and whatever? Or was it still relatively new so we didn't really—

CEP: It was relatively new, but there was some literature and there were very few pests. It was known that did extremely well on fertile, well-drained soils. At the time in '69 when I started at Jari, the plantations that he had started in Costa Rica to guarantee seed source were probably five years old, and the growth there was fabulous. Of course, it was on banana-type lands. The conditions in Costa Rica were such that the soils were very fertile but not well-drained, and so they had gone in and ditched. They kind of gridded the plantation and ditched it to provide better drainage. And the growth was absolutely fabulous.

HKS: They knew enough to drain it.

CEP: Right.

HKS: So it wasn't guess work. There was some basic plan.

CEP: That's right.

HKS: If you want to grow gmelina, you have to do these kinds of things.

CEP: The literature from stands in Burma—it doesn't occur in pure stands—but the literature from Burma simply said it prefers fertile, well-drained soils. So the soils in Costa Rica were fertile but not well-drained. And so they drained it.

At this same time, at the citrus plantation in Panama, probably in '63 or '64, when he first started thinking about it, they planted a block of maybe an acre or two of gmelina. It was good soils, and the growth there was just absolutely fabulous. So he was making the assumption that he would get the kind of growth in the Amazon as he got in Panama and Costa Rica.

It wasn't just me. Every forester that we had with any kind of experience was convinced that it would not provide that kind of growth on our sandy, sedimentary soils. This will go back as far as probably when Bassett McGuire first visited. He had the same feeling. When I arrived, a road had been punched north a good ways to the better soils that I previously described, and several hundred acres cleared so that they could start planting on those better soils to show the difference in growth between sandy soils and clay soils.

Probably everyone with any training or background whatsoever had the same fears that we did. We were under pressure to plant the areas closest to the port first, simply from the standpoint of logging cost and everything else. So the bulk of the areas planted in the earlier years were on sedimentary soils, where all of us were uncomfortable with gmelina large-scale.

The one comfort we had was that we were on a five- to eight-year rotation, depending on the soil. All of us were convinced that we would get at least one rotation on those sandy soils, and by that time we would know enough to replant the second rotation on those lands with some other species, whether that would be pine or eucalyptus or whatever.

If it had been a situation where we thought we would not even get a first rotation, then the fight would have been much more vigorous than what it was. But under the assumption that we'd get one rotation—which we did—and that he would let us change when we knew what to change to, we proceeded.

He essentially forbade us from doing anything with any other species for the first year or two. He knew of our concerns. He was afraid that if he said, "Okay, do what you need to do," that the initial effort on gmelina would be watered down. He said, "After a couple of years, when we see how it's going, we'll do what we need to do."

I wasn't willing to wait. We obtained *Pinus caribaea* seed, and I went into backsides of plantation blocks that he would never see, except from the air after it was too late, and we put in several one-acre plantings of *caribaea*. There was one in particular that was just about a kilometer off the road between the town and the port. When it was probably eighteen months old, I took him to it. It was absolutely beautiful. The color was as dark a green as you could ever want, and the growth more spectacular than any pine I had ever seen at that time in my life. We drove up, and he said, "What's this?" I said, "It's pine."

HKS: This is an acre.

CEP: About an acre.

HKS: Okay.

CEP: He said, "How much have you planted?" I said, "This, and a few more just like it." He said, "What's wrong with you? You ought to be planting ten thousand acres of this."

HKS/CEP: [mutual laughter]

CEP: That was about July. The following January, which was six months later, we planted ten thousand acres.

HKS: You could get a plantable pine seedling in six months?

CEP: Yes.

HKS: That's incredible. I don't know what it takes. I mean, Douglas-fir is like two years or something.

CEP: Right. You don't have a dormant season. I didn't dare plant bare-rooted and all my experience in planting was bare-rooted, in the winter.

I went to what was then St. Regis in Florida to a milk carton plant, and I designed a small, about a one-and-a-half by one-and-a-half inch square, six inches long, folding carton with holes punched in various positions along all four sides and the bottom. We used those as containers. So we built frames. We'd fold these cartons, build a frame, fill it, and then fill that flat with a premixed soil or growing medium, and put the seed in that, and then all of that in nurture beds. So if you drove by and looked at the nursery, it looked like a regular pine seedling nursery. But they were all in containers. So in planting season, you didn't disturb the seedlings at all. You'd just go out and dig a hole, drop this container in, and fill in the sides, and the roots came out through the prepunched holes.

From the decision in July until January—we planted in January, February and March—but by the end of March we had ten thousand acres of these.

HKS: Were you still planting gmelina?

CEP: Yes. Ten thousand acres wasn't enough to cover all of the available sandy soils, so what we did, our soils group mapped the areas that had been cleared, and those that were more sandy than others, we put into pine. That was 1973. Each year after that, the percentage of sandy soils that went to gmelina declined until about in '76 all of the gmelina was being planted north on soils formed in place, and all of the sandy soils were going to pine.

HKS: Was there adequate literature that allowed you to pick *caribaea* over *radiata*, over some other species? And was *caribaea* the obvious choice?

CEP: *Caribaea* was the obvious choice because here we're dealing with pure low-land tropics. *Radiata* is not a tropical. It's grown in some regions in, say, the sub-tropics. *Caribaea* is not truly low-land tropics, but it's nearest in its native range in Central America. It comes closer to the pure tropics than any other species.

HKS: Is that the species used in southern Brazil? *Caribaea?*

CEP: No. In southern Brazil, it's mainly slash and loblolly.

HKS: Okay.

CEP: Here again, southern Brazil is temperate. In the northern portion of the range, where pines have been planted, you can consider some of it to be sub-tropics. But most of the pine is in a temperate location.

HKS: I'm impressed that you can get enough seed to grow seedlings for ten thousand acres, or almost ten thousand acres, within a short period. There's a seed source there?

CEP: The first seed for that first year was a purchase from Resource Management in Birmingham, Alabama, who I had had business dealing with before. There's a bulletin by A. Lamb. It's an Oxford publication on *caribaea*. And being a pine guy before I went to Brazil, and knowing Lamb personally, I knew that publication in and out. We knew the location of all *caribaea* plantations at that time.

At the same time as we bought the seed from Resource Management, I put together a team and went with them and got them started on seed collection in Central America, and we collected enough seed that season for several years' planting. Plus we did a standard collection for providence testing. We established, both at Jari and another location, providence tests to determine which source within the range of *caribaea* was best for the Jari situation.

Maybe this is the point to say also that Weyerhaeuser at that time had an interest in—they had a—I don't know how to describe it. Let's say they had a corporate feeling that at some stage, somewhere in the world, next year or thirty years in the future or sometime, they would need and establish large-scale pine plantations. So Weyerhaeuser was as interested in having a seed source of improved *caribaea* as we were.

From the base at Jari, we started all kinds of things. When you'd have an idea that we couldn't control, the Jari base was responsible for carrying out whatever activities. So we were involved in mining exploration and everything you could imagine. In fact, the present-day ALCAN bauxite program in Brazil was initially our program. We developed that mine, and then it was sold to ALCAN.

In the State of Minas Gerais Ludwig, along with Antunes. Antunes is the individual's name. The company name is a conglomerate called ICOMI. Anyway, Ludwig, with Antunes started a project in the state of Minas Gerais. This was back in the early '70s. There was the oil embargo, the high interest rates, the economy was just absolutely slaughtered in Brazil. People here thought it was bad, but there, at that time, Brazil imported ninety-two percent of their carbon fuels requirements.

HKS: Wow.

CEP: They went out into this vast open place, like going to Oklahoma with nobody there, and started an alcohol project based on sugar cane and manioc. That gave us a location, a sub-tropical location, to establish provenance tests and a seed orchard in a situation where there would be no outside contamination from pollen.

That seed orchard was a joint Jari-Weyerhaeuser program. Weyerhaeuser had a geneticist—his last name was Dykstra—who I believe had been in their project in Indonesia. We had already started it and got it going, but they assigned Dykstra to that project. It was carried out as per plan. Exists today. Weyerhaeuser has pulled out, and I do not know the details, but as far as I know, they have pulled out but maintain an agreement whereby if they need seed, they have the right of first refusal to purchase seed at market price. Some combination. I don't know what the details are. So for anyone desiring to establish large-scale *caribaea* plantations in the tropics, the seed is there.

I mentioned that in about '74 we started planting in the savannas, in the Territory of Amapá, and that initial effort was—ICOMI had a manganese mining operation there. They exported manganese, and they could begin to see the end of the reserve. Antunes asked Ludwig, "What am I going to do when I run out of manganese, and we have this tremendous infrastructure? A when I run out of manganese, and we have this tremendous infrastructure? A town: hospitals, schools, everything. Just like we did at Jari. Railroads from the mine to the port, port facilities, the whole deal."

I don't have time to go into the details, but Ludwig had a great affinity for Antunes for some specific reasons, and so any time Ludwig could help him, he wanted to. For that reason, Jari started doing developmental work for large-scale pine plantations in savannas.

We did ripping to break up the lateritic layer. We plowed, disked, planted in the grass every combination you could think of. And on the Jari property we also had many savannas, in isolated places, and we would just put a crew on a helicopter and drop into a savanna with planting bars and containers of seedlings over your shoulder, and the helicopter would sit there and they'd lay off a plot and plant four hundred trees and get back on the helicopter and leave.

HKS: You could do that sort of stuff in part because in the early stages there was no cost accounting. There was no cost-effectiveness to worry about. He wanted you to work on other projects. It didn't affect your success at all.

CEP: That's right. He was interested in overall development of the property, regardless of what it was. It didn't have to be gmelina. It could have been squash, so to speak. At another location on the property—which is a total different story—we had a major rice program, with the research and developmental work under contract to the International Rice Institute, which was an arm of the Rockefeller Foundation.

He was interested in total possibilities for that property. He had a vision, and he had a place to start. But past that, it was no holds barred. And so we began doing that developmental work for pine. I did it before he ever knew. I don't know the number, but this group now has planted probably a hundred and seventy-five thousand acres of pine. Some of it has been hauled to Jari, and barged to Jari and pulped. That is a resource available, and there are literally millions of acres of savanna now with this proven project and proven growth rates and proven better growth in successive rotations, for somebody to go into savannas large-scale. Now a Weyerhaeuser, a GP, a large corporate organization could now go in justifiably and not be slaughtered. The bulk of the unknowns had been answered.

HKS: What's happening to gmelina all this time?

CEP: In an endeavor like that, in retrospect, when you have twenty-twenty vision, we made mistakes. They weren't mistakes, I guess, because no one knew. Didn't know any better. So they were mistakes only in terms of twenty-twenty hindsight.

The first step was gmelina. We added pine by brute force, not by any data or information or anything else. But from day one, our research group started species testing. So on every soil type we had plantings of native trees—I mean, we'd see para-para, which is a pioneer species, and it's lightweight, uniform type fiber, grows beautifully straight. We'd collect seed and we'd grow para-para.

Literally any species that appeared to have promise under plantation culture, we planted. Any species that had proven to be useful anywhere in the tropics anywhere in the world that we could identify, we found and planted. We didn't go out and advertise. But all of those plantings were done and data kept.

We also assumed that there was one or more eucalyptus species that would be suitable on the sandier soils. Unfortunately, there are many eucalyptus species, but very few that are truly tropical. There are only about four or five species that are truly tropical or low enough sub-tropics that you could say, "Well, I'm going to call it tropical."

HKS: This would be in Queensland where you looked.

CEP: Right. So we planted plots of every eucalypt we could find, including the ones being used in central Brazil in large-scale plantings. Eucalyptus had been planted for a long time in Brazil. Much of the smelting activity in Brazil is based upon heat from wood charcoal. The original forest was cut for charcoal, and they came behind that and started planting plantations of whatever species, and it has turned out that eucalyptus is one of the better. So there are large-scale eucalyptus plantations, short rotation, that are used for charcoal for the steel industry.

There's long-term experience on species in central Brazil. At the same time Jari was being developed, there were large-scale plantations coming onstream for pulp. There are many large-scale eucalyptus plantations in central Brazil that are managed like Iowa cornfields. I mean, it's plowed, fertilized, weeded, irrigated, and the growth is just absolutely phenomenal.

There is a fair amount of literature on eucalyptus also. Because of the freedom we had, anybody we found that had successfully done something with eucalyptus somewhere in the world, we tagged him and brought him to Jari and said, "What do we do?"

HKS: Was there an underlying assumption that you'd have trouble with pests because you were in the tropics? You wouldn't necessarily know what they were in advance, but the literature brings it out a lot about the ants and the Mirex. Ant control could be major.

CEP: Right.

HKS: Henry Ford lost his shirt in Fordlandia in the '20s because of the rubber plantation and disease, so I didn't know what your expectations were.

CEP: My assumption is that a tropical tree that is not under stress does not have any more risk for pests than a temperate tree that is where it is supposed to be.

HKS: So a plantation managed with single species with all the risk of a single species, you weren't concerned about that, more than you would be anywhere else.

CEP: No, and that's proven to be the case. Now, what little I know about Fordlandia, Ford, as everybody knows, was a very strong personality. In digging back through the literature and listening to people, my analysis of what happened is that the scientific people he had were good people. They knew to maintain a broad genetic base, and they were doing that. They came up with a few clones that were absolutely fabulous, and with his personality he said, "You're going to grow those." I mean, rubber production is high, they're early producers, whatever the traits were, "and this is what you're going to do." Personality-wise, they didn't have people like maybe we had at Jari. They did what Henry Ford said. They had no genetic base. They had very few clones, and when something hit, it was over.

Our program at Jari, for pine, as an example. Every origin of pine, every environmental niche, every extreme of elevation—on top of a mountain, the poorest soils, growing in a swamp, the southern end of the range, the northern end of the range—every extreme that the species occurred in was planted at Jari. And maintained in blocks. The records are kept in such a way that if something shows up at Jari, say in pine.... Let's say that one region of the natural range is susceptible to something at Jari. On a large-scale planting, the seed is mixed and planted at random, and so yes, an occasional tree that's not adapted there and a bug gets those out. But it doesn't wipe out your plantation. Then you leave that particular source out of the next planting.

HKS: I think it was the articles by Fearnside and Rankin. Are you familiar with their work? They're trained in ecology and were observers. This was about the time you left, or after you left. They went inside, just describing the ecosystem, and published a sequence of articles. Gmelina, if I read it correctly and remember what they wrote—there was a canker that made gmelina not work. Now, is that gmelina growing under stress, as opposed to gmelina growing in the soils where it was adapted?

CEP: Well, there's two things. Early on, we were cutting gmelina, let's say at the end of the first rotation. Cut it, and it resprouts. Then, after a few months, you go back, and the dominant sprout has expressed itself, and you knock the others down. Under those conditions, and in sandy soils, where it's not adapted, you have the problem.

HKS: With canker.

CEP: Right. We had things show up. We had an insect, a caterpillar show up that none of us had ever seen before. On a given soil type it showed up one day and defoliated the place. We thought, "Oh, my goodness." Rather than panic and go out and spray and kill everything, I said, "We're just going to watch it." It disappeared.

I mean, you take southern pine in east Texas today. That species was born there, so to speak. It's in its native range. It's in its home, and it's being attacked vigorously by southern pine beetle. Now, who made that mistake?

You can use examples. Take fusiform rust in southern pine, across the range. Whose fault is that? So what tends to happen is that when anything shows up in a plantation, regardless of severity, it's because it's a monoculture, and someone should have planted something different or whatever. But any species out of where it is comfortable, so to speak, is more susceptible to both insect and pathological problems. Of course, that was one of the reasons for us pushing so hard to get the right species on the right site.

HKS: So you didn't have any particular concern about monoculture, about being wiped out by something. Because the literature makes a big deal about—

CEP: Oh, yes. Oh, everybody that came through just wailed and moaned.

HKS: Well, you learned that when you were a junior in college.

CEP: That's right. These guys that came through, they would say, "You can't do this. It's a monoculture." I'd say, "Well, get in the jeep. Let's go." We'd

HKS: What do you mean by "cleaning?"

CEP: The jungle comes roaring back.

HKS: It's the jungle you're cleaning, not the pines.

CEP: Yes. The biggest single cost at Jari was not land-clearing, not building roads, nothing other than maintaining in good growing conditions the plants that we established. When you go in and a crew goes through and they knock down the competing vegetation with machetes, it looks like a pure, quote, "monoculture." And it's pretty. But when—

HKS: Now, that's gmelina, and here's your son there. [showing photo]

CEP: Right.

HKS: Is that a fairly close canopy stand? So even if the stands grew not to maturity but fairly full, there's still enough sunlight getting through that you have the maintenance problem.

CEP: Absolutely. Gmelina, as you see in this picture, is deciduous. In the dry season, it drops all its leaves. So you fly over it, and it's all dead. During this few months period, when it has no leaves, sunlight is on the ground, just like it would be on the forest floor in the U.S. in the wintertime. Except that the plants that are not deciduous then are getting full sunlight. Obviously, a picture such as this, a crew had just gone through, and it has been cleaned.

But I would take these people. I would ask them for a definition of monoculture, and then I would take them to a stand that needed cleaning and say, "Does this fit your definition of a monoculture?" They would look up and say, "Yes." I'd say, "Well, lower your eyes and tell me what these other hundreds of species are doing here if it's a monoculture." It would kind of set them back.

In the lowland tropics you have a species of preference. But it's certainly not a monoculture. I've never understood the phobia over monoculture, because if you fly to northern Finland, or to some places in northern Canada. In northern Europe you can go just miles and miles and miles and miles of pure *Pinus sylvestrus*. Is that not a monoculture?

HKS: It sure is.

CEP: So, "monoculture" in and of itself is not a nasty word. It all boils down to management. Now, to me, monoculture, in the absence of management, in the absence of genetic diversity, in the species you're dealing with, in all likelihood will lead to disaster. It's just like clearing tropical soils and doing everything wrong. Yes, it will lead to disaster. But it doesn't have to.

HKS: So good management is the issue, not monoculture or anything else.

CEP: Right. If monoculture is bad, how do these natural monoculture stands survive all the onslaughts that they talk about?

HKS: Okay. I've asked you the question. You've answered it.

CEP: [laughing]

HKS: I don't want you to challenge all of it, but it came up everywhere!

CEP: Everywhere.

HKS: One said gmelina rotation has changed from six to five to four because of the canker. It's just generalized statements.

CEP: It depends on the site. Of course, when these people were there was when there was still extensive stands of gmelina off site.

HKS: Okay.

CEP: So, back to one of your questions. Today, at Jari, twenty-five years later, it's pine, *Pinus caribaea*, gmelina on the very best soils, and eucalyptus on the in-between.

HKS: No native species.

CEP: No. No. And we never even got close with a native species.

HKS: All kinds of problems? Or what?

CEP: Many of them simply are not adapted to plantation conditions. There are many U.S. species like trying to grow walnut in plantation conditions. It's very difficult for it to work. We would find a species that looked good, we were interested in. We'd do pulping studies on it, and it wouldn't be any good. So for whatever combination of reasons, we just never came up with a native species that would compete with pine, eucalyptus, or gmelina.

HKS: You had a rather admirable research program there. You started answering the questions you had on site. Did your scientists publish in the technical literature, *Forest Science* or whatever?

CEP: Very little.

HKS: You had a rather admirable research program there. You started answering the questions you had on site. Did your scientists publish in the technical literature, *Forest Science* or whatever?

CEP: Very little.

HKS: That was a personal choice? I dropped out of forestry twenty-five years ago—been in history since—so I don't know what's been going on. I don't read the technical literature except maybe the *Journal of Forestry*. I scan that. Using my technical knowledge of the late '50s, early '60s, this is new information.

CEP: I say this in retrospect, and it wasn't a conscious decision that I made. Some of us occasionally went to meetings and made a general presentation, but number one, we were in a struggle for survival. Everyone there worked thirty-six hours a day, ten days a week. And loved it. There was so much demand for information of what to do next and watching what we had already done that there was simply no time for publication.

## Ludwig's Consultants

CEP: The other thing that was probably subconscious in most of us was that we were under constant harassment and bombardment for being pioneers that we had it up to the gills with propeller-tie scientists and consultants. Ludwig had no personal limit on how much he spent on consultants to run around all over the world. I mean, you would have a truck transportation expert from somewhere show up to advise us on forestry. You would have someone from an accounting firm show up.

One of Ludwig's bad traits was that he was in such a hurry that he did not have time, and he didn't have the personality to choose the right person for the right job. He simply put someone there, and if they worked, fine. If they didn't, they disappeared. That was the selection process.

To give you an example: We had been in the field one day. We came back to the office, and I was walking in front of him, and something that he saw in the field that he didn't like hit him, and he had forgotten that he hadn't said anything to anybody about it. He stopped the first guy that was nearest to him, told him what he saw, and told him to be sure that that never occurred again. The guy didn't speak English. He was a janitor. In his mind, he resolved the problem because he told someone to take care of it. Now, fortunately, I stepped back and heard what he said, and I took care of the problem. But if I had not heard it, when he came back next time and he saw it again, you'd be chewed out for not solving the problem.

HKS: Chain of command. That really wasn't one of the things he was concerned about.

CEP: No. No. And realizing many of these things about himself, one of his protective mechanisms was to have a steady flow of people in and out, whether they knew anything about the subject or not. Going through and going back to New York and telling him, "The trees are dying." "The trees are living." In one case, the guy went back, and his idea of a tree was that it's beautiful and green and it's big. And he came and he commandeered a vehicle, which all of them did. He didn't know where to go, so he kind of followed traffic, and he got out into some of the areas that had just been planted with gmelina stumps. Well, there were no trees there. There was no green. He went back and told Ludwig that everything had died. [laughter] So we get a call. He says, "Surely this is not true." But this guy comes back and says all the trees are dying. I said, "Well, I hadn't noticed." [chuckling]

HKS: How would Ludwig find these people? They'd have some sort of credentials or track record or something for him to bother talking to them or hiring or retaining—

CEP: Not necessarily.

HKS: You mean if I had sent a proposal to him someplace, he might say, "You're a historian. Go down and study the history of this. We need this."

CEP: Yes. Absolutely. We had one guy who came down as a consultant on trucks. His name was Larson. He was from Portland. Ludwig met him on a boat somewhere. Ludwig talked to people. He said, "What are you doing?" The guy said, "Well, I'm old and I just sold out a fleet of trucks I had, so I'm spending my time redoing my boat." Ludwig said, "You know, I'm putting together a project in Brazil, and I've got to have hundreds of trucks. You need to go down there. Would you go down there for me?" "Well, I just retired. I guess I would."

HKS: [chuckling]

CEP: For three years we had this guy who is our truck, transportation, specialist. I don't mean one guy occasionally, I mean a continual flow of people like that that you have to learn to deal with, live with, contend with, whatever the word is, and get your job done at the same time.

Zeb White was a consultant. He was there several times, and he was in the class of the comforting type of individual. There were times that we used him specifically when we were making World Bank presentations in dealing with forestry experts in the World Bank who were not forestry experts at all.

HKS: Now, you weren't trying to get money from World Bank. You were just reporting to them on tropical—

CEP: No. We were getting development money. We got to the stage with World Bank on the hydroproject and on the newsprint project, mechanical pulp project, that the only thing that remained was approval on the Brazilian side, and we were in. We had major battles with World Bank, just on terms such as mean annual increment. Now, everybody in forestry knows what mean annual increment is. We had our World Bank presentation completed, done, and a forestry expert came down and looked at three-year-old pine plantations. I don't remember the numbers, but let's say that our mean annual increment was three times greater than what you'd have in the South, which it is. And that's a lot of volume. He looked at three-year-old trees, and there was not that much volume there. So we couldn't know what we were talking about.

Well, a tree has to live its first year, and it has zero volume. It has to live its second year, and it has zero measurable volume, maybe. So in the States, an inventory system only starts once a stand is old enough to have a measurable volume. So a ten-year-old stand in the States has very little measurable volume. So mean annual increment of that actual stand is zero.

But your average annual growth is based upon a rotation. You have to have a first year, even though there's zero volume. If you go out and measure a three-year-old plantation, you have zero growth. And he did that. He went back and reported to his bosses that our projections had to be horribly in error, because he looked at plantations that had very little volume.

We would use consultants of our choice who had the stature and capability of standing up and explaining technical things to non-foresters. Tommy Thompson is in that kind of category in that he and Ludwig hit it off and got along extremely well. We were interested in development of the natural resource, and we were interested in milling capacity at Jari to match the volume that we thought we could sell on the world market.

We were not interested in milling capacity to mill everything that existed there, because there's a world of difference between what you can mill and market, and the total resource available. So Tommy helped us plan and design facilities along that track, and helped us deal with Ludwig on that subject when Ludwig had guys battering around the edges to build a monster.

In the end, the monster got built. But that was one of my major final battles.

HKS: That's the big sawmill you're talking about.

CEP: Yes. It's one of the big log mills in the Amazon that never functioned. One of many.

HKS: So the consultants of most value to you were those you selected. You had a specific problem, and needed specific sort of advice.

CEP: Had a specific problem. Had an innate fear that we might not be going down the right path and need somebody completely disconnected from the daily fray of battle to freely look over our shoulders and say, "Rah, rah. Keep going. Turn left." Or whatever. And in most cases it was never a formal report written. It was a hand-holding session.

HKS: Yes. Tommy told me he made sixty-two trips to Brazil or some large number. This was after he left GP he became your consultant from time to time.

CEP: Right. He was the only one that was our consultant and Ludwig's consultant.

HKS/CEP: [mutual laughter]

CEP: There were times he was there when Ludwig would call him and say, "Please go down." There were times he was there that Ludwig may not even have known he was there, unless Tommy called him and told him, because we didn't.

## The Jari Railroad

HKS: That anecdote reminds me of something I was reading that Ludwig had these preferences for certain things, and one is that he preferred—he wanted a railroad. Even when trucks would have been more economical to build the road as opposed to the railroad. I don't know if it's true. But first of all, he wanted the railroad. There was a railroad there, right?

CEP: Yes. We built a railroad.

HKS: It was, in your opinion, the proper thing in terms of the overall infrastructure needs of the project? Or would a good truck road have been just as good?

CEP: Under tropical conditions, it would be hard to conceive of a truck road that would withstand the traffic. You get into the rainy season, when it rains every day, and you have literally hundreds of trucks daily on the main road system. The only possible road that would withstand it would have been a concrete road. The rail system was a joint effort, economic study, between ourselves and Weyerhaeuser. It's proven to be correct.

HKS: So it wasn't because one of these consultants came through and sold Ludwig on the idea of this.

CEP: No.

HKS: This was a good idea.

CEP: With Weyerhaeuser we had such a relationship that if we needed two or three transportation specialists, all we had to do was call. We had Weyerhaeuser plantation people, economic people, transportation people, facilities planning people living with us for over a span of at least three years.

HKS: We had GP in Brazil, McMillan-Bloedel is there. I don't know who else was there. But Weyerhaeuser somehow could see the benefit to the company and collaborated on certain elements of it.

CEP: The overall objective was that there were joint venture negotiations between ourselves and Weyerhaeuser. At that time the objective was that Weyerhaeuser would come in on a fifty-fifty basis. We were at the stage where we had proven the viability biologically, and we were needing facilities planning people, engineering, mill start-up capability—all those kinds of things to move us from a developmental biological project to production.

HKS: Weyerhaeuser's Bill Johnson. Was he the primary figure in this?

CEP: He was there, and he was probably involved more with railroad than anything else. My feeling about Bill Johnson was that he was, at that stage, already an old-timer. He was a blood-and-guts, salty production kind of guy that George Weyerhaeuser liked. I don't know whether this happened across the board or not—but what I saw was that George Weyerhaeuser listened to the scientists and the management people, got all of the data necessary to make a decision, and then he would ask Bill Johnson, "Does this make any sense?" Bill Johnson had the ability to cut through all of the pages and the extra verbiage and say, "It sure ought to work," or "No, we need to forget it."

Bill Johnson did not have a specific direct responsibility. He was in and out occasionally. In fact, we used him in the same way. If we and the Weyerhaeuser technical guys were there and we came up with a conclusion, and he came down and would say, "This makes sense."

Anyway, the railroad was a planned deal. To this day, I'm still convinced that with the traffic involved there's no other logical way.

HKS: Roadbed maintenance I guess would be the issue there. I was also thinking about maintenance problems. It's one thing to repair a diesel engine on a truck, but to repair a locomotive is a whole layer of skills and harder to find.

CEP: Right. But see, Antunes had a railroad just east of us, in the Territory of Amapá, and it had been there probably twenty-five years when we built ours. So people to maintain locomotives were just across the river, so to speak.

But when you imagine an eight hundred and fifty ton a day pulpmill and all of the fuel that would be required to provide the power, the volume is tremendous, and I think I'd be safe in saying there's not a single highway anywhere in the U.S. that carries that kind of industrial traffic.

The port was really the only place it could be. Just above where we put in the port, the river became shallow. So we were at the upper limit of the port. The port, if you flew over it, you wouldn't call it a peninsula, but when you look at the Jari River and the small river that runs into the Jari at the port side, and turns and runs around to the west side, southwest, in swampland, in reality, from the town to the port is a peninsula of high ground. It's really the only place the port could be. So you have a stretch of road that carries one hundred percent of the traffic. The town of Monte Dourado was in the only logical place.

The kaolin refinery was at the port. Pulpmill at the port. Because of the isolation, all of the chemical recovery was at the port, and some chemical processing. Power plant at the port. So if you can imagine supplying all the raw material down a road and having all of the work traffic of all the employees down that same road, you literally would have had to have had a six-lane concrete highway. So yes, it's easy for somebody to fly in and say, "What are they doing with a railroad here?"

We didn't want a railroad. We tried in layout and every possible way to prevent it. Ludwig didn't want a railroad. Until we started looking at numbers on traffic. We ended up with a railroad when, to my knowledge, he never wanted a railroad.

HKS: At that time, you were planning for a second mill. You needed that capacity.

CEP: That's right. That's right. At that time, we were planning a newsprint facility. We were planning a dam on the Jari, which is now, presumably, going to be built. I mean, the concentration of traffic is just horrendous. Also, the kaolin facility has been a good investment and essentially profitable since initiation. The kaolin comes from east, just across the river. It's on a high plateau. It's put into a suspension, centrifuged to get the larger particles out, and then gravity-flowed to the refinery. That reserve will not last forever. There's another reserve just to the west on the plateau. And there's a railroad spur to that location. So when you look at overall development, there's just no logical other alternative.

Almost regardless of what the subject is, you can see in the literature things that somebody flew over, spent thirty minutes, and became an expert.

Things happen so rapidly. I took a group of about forty people to Jari three years ago, and a lot of the people that we trained were still there, and we had a fabulous reunion. But management did not know me from Adam, and I didn't say anything. We went on the tour, and they took us into pine plantations and took us here and there. On that trip, among many other things, I discovered that present management is the one that started planting pine, and present management was doing a lot of things that they had developed and discovered, and anyone very astute could have looked at the plantation and seen that it was twelve years old, maybe, and said, "How long have you been here?" And known the difference. That is not atypical of what happens in any kind of a management change.

HKS: You said something earlier, and I'd like to follow up on a little bit. That eventually you had more people in research than you had in management or some managerial positions. I don't know what kind of financial statement that Ludwig received from time to time, but he must have noticed that.

CEP: Well, that statement was related to forestry only.

HKS: Forestry only.

CEP: Yes. We had a soils group, we had a tree improvement group, we had an inventory group, we had a native forest group, because the intent from the beginning was as much as possible to utilize the natural resource that was there. We just never were able to succeed at it. Of course, no one else has, either. We didn't do any better than anyone else.

HKS: Native woods were used for fuel, weren't they?

CEP: Used for fuel.

HKS: But that's it.

CEP: That's it. Of course, that statement has to be flavored by if you were in southern pine region and your name is Union Camp, you would have a group doing continuous inventory work so that at all times you know what your position is on inventory. With us, we had that group, but it was developmental, and they were in the applied research group, rather than in the forestry operating. Because we were learning what to do and how to do it.

If you took the normal guys that would be in an industrial forestry department and put the inventory guys on the operating side, and go through the organization in that manner, that statement may not have been true. But for us, because we were learning, if you were not directly involved in clearing, land prep, nursery planting, plantation maintenance, then you were doing developmental work. Now, we did not, for internal political reasons, we never had a group that we labeled "applied research." At the time, we did no research, as far as our organization was concerned, because it would have been seen as, and some accountant would have tagged it, as research. We were supposed to be busy establishing plantations. And so a geneticist there was doing applied research to help guide us to know which direction to go, but we certainly did not give him a research title.

In fact, the guy running the soils lab we sent back to graduate school. He got his doctorate in soils at the University of Maine. He was a scientist, probably in the truest sense. He was Brazilian. He was certainly doing scientific work, but we never called him a soils scientist. He was involved in site selection. He had to do all of this background scientific work in order to select the site.

## Hiring Brazilians

HKS: Sure. Were many of your scientists Brazilians?

CEP: In the early stages, no. One of the unusual comments is that Brazilians are very proud of their country, and an educated, upward-moving individual, scientist, wants to be where the action is. Well, the action is in the south. And, in general, Brazilians have much more fear of the Amazon, and all of the snakes and mosquitoes and alligators and whatever, than any other people on earth. So to attract a Brazilian professional to go from where the action is to the isolated nothingness is extremely difficult.

It's the same mentality in our New York office, when I said, "I've had it with this place. I'm moving to Oklahoma." The New Yorkers there couldn't believe that anyone would voluntarily leave New York. I mean, yes, you could be banned to Oklahoma, like people are sent to Siberia. But surely no one would voluntarily go.

HKS: I understand that. Did you have difficulty attracting...

CEP: You could hire any nationality on the face of the earth easier than you could hire a Brazilian.

HKS: Were most of your scientists from the U.S.?

CEP: Yes.

HKS: Did you have trouble attracting them?

CEP: No.

HKS: They were hired for a two-year stint, bring their family.

CEP: No. I never hired anybody except for forever. They might not last but three weeks, but the intent was forever, because our program was forever. In an unknown situation, if a guy stays two years, it's a waste of his time and ours. It takes at least that long to adapt to a new language, new culture, and new biological world before one can become productive.

In an extreme case, we had a management guy who was not succeeding in carrying out his job responsibilities, and I couldn't understand why, because in talking to him, he knew what to do, he knew when to do it, he left at six o'clock every morning, he came in after dark. Nothing was happening. He was technically qualified to do the job. I showed up and went to work with him one day. He was hollering at some of his supervisors, in frustration. I said, "What's the deal?" He said, "I've been here six months, and not a single one of these supervisors speaks English yet." I didn't say a word. I went back. I sent a telegram to Belém, got a ticket for him and his family, and put them on the next airplane back.

Because one of the requirements that I placed on anybody I hired was that within three months they didn't have to be fluent in the language, but they had to be able to get their job done in the native language. If they didn't, they were automatically gone.

HKS: But you had language training there.

CEP: Yes. And you'd be amazed. This is somewhat unfair in that I don't have difficulty with languages—but you can learn the word for shovel. You can learn the word for tree. You can learn the word for snake. You can learn a hundred words, and by saying those hundred words and gesturing and showing, I mean, these people learn, the natives there learn very rapidly. They want to learn. They want to do better. They are determined to show their boss that they can do a good job. So getting a job done in the native language is not all that tough. You have to want to.

What we did was initially bring in people from the outside, and then Brazil has excellent technical schools. You go through high school, and then you have two years of technical training, in forestry or agriculture or engineering. You take a basic physics course, a basic biology course. Enough to at least get an introduction. Then you go to work.

We found that these technical people, young, aggressive, were readily trainable, and we had extremely good experience with young foresters. We never had good luck with getting a forester with five years experience with an industrial organization in southern Brazil. But taking guys fresh out of school, first job, they were hungry. Had to have a job. Willing to go anywhere. Had to go to the Amazon, because that was the only job there was. But once they were there and found out how terrible it was not, and got wrapped up with the challenge, turned into just absolutely fabulous guys. I'd put them up against foresters from any school in the world.

It was that core, that type of core of technicians and young foresters that we trained. We had the freedom to send them to see and look and learn anywhere in the world that there was something for them to see. We built the foundation with those kinds of guys, and they're still there today. They disappeared in the change, when we sold, but management learned and had to go find them and bring them back. They're there today.

The guy that for the last eight or ten years has been responsible for all of the harvesting program was the guy that I trained to run the nursery. So all of those guys are in management positions. As we were able to train those people, then we had fewer and fewer and fewer expatriates, to the point that when we left, the only expatriates we had was top management and scientists in specific fields where they simply were not available in Brazil and the kinds of scientific work that you can't take a guy with a B.S. and train him.

## Hiring Americans

HKS: A thought I had was that many Ph.D. scientists that I know of, in this country, are somehow tied to the academic system, and you have a tenured position. It's one thing to ask your dean for a leave of absence for a year or two, and something else to sign on indefinitely. For them to give it up, I thought it may be difficult for you to have attracted people because of that.

CEP: But these kind of people would not have been qualified for Jari. In their minds they're probably qualified to do most anything. But they would not have been qualified to work at Jari and would not have survived.

HKS: Someone like John Welker.

CEP: He's with Mead Corporation. He walked out of Yale with his masters in economics and came straight to us. So he had all the background, and we trained him on site. He's an avid reader, so he continued his education. We'd give him a job that he had never done before in his life, and he had to do it. So, by talking and reading and discussing, he'd succeed.

It takes a pioneer mentality to be involved in something like this. One of the most difficult things for most everyone involved in the developmental stages of Jari is that rarely are the what I call the blood-and-gut pioneers—whether they be management or scientists—rarely are they qualified to remain and operate a Jari once it is established and stabilized and an economic, producing unit.

To walk away from your child that you have devoted your life to, day and night, and seen grow and develop and function, and to learn that you're not qualified to stay and participate is one of the most traumatic experiences. Unfortunately, some guys haven't survived it.

HKS: Let me ask the question one other way: Were most of the people who we're calling scientists Ph.D. scientists, or were they people like Welker, with a basic education and a lot of potential? You said McGuire made the mistake of saying you had to have a Ph.D.

CEP: Right.

HKS: Did you learn from that mistake?

CEP: We had many Ph.D.s over the period. In a situation such as that, I had several criteria in hiring. One was that the interview of the wife was more critical than the man.

HKS: I can understand that.

CEP: Because I knew that almost any professional would get so wrapped up in the challenge and in the activity that he would be satisfied. So if a wife did not have a burning desire to paint, to sew, to do ceramics, to be a volunteer in the hospital, to be a volunteer teacher—if she didn't have a burning desire to contribute to community.... [chuckling] I'll be shot for saying this. But through experience I learned that any wife who had to have her hair fixed once a week would not survive.

HKS: It's probably much more common today than it would be fifteen years ago, but you can hire a husband and wife team that are both trained in science. You could hire them both.

CEP: Right. Then it would have been very difficult.

HKS: That would explain why the lack of publications and so forth. They're just different. They're in a different universe. You didn't pull them out of the university setting.

## Freedom to Make Mistakes

CEP: That's right. That's right. Essentially, what would occur is that a guy never dreamed of having so much freedom, and so if he had gone to work for an industry, in an industry you get pigeon-holed. I mean, you're an inventory specialist and you may become the best inventory specialist in the U.S., but you're an inventory specialist.

Unfortunately, in our country we pigeon-hole everybody, professionally and otherwise. Here, you hire a guy, and you simply say, "Your responsibility is establishing a silvicultural program for this project. Go get it. Do it." And the guy's mouth drops open, and he shudders, but that's what the job was.

So when the individual learns that latitude that he has to dream and that he can make honest, professional mistakes—and in an unknown situation, I guarantee you will make mistakes. If you don't have the ability to look back and say, "Hey, we scheduled pruning these trees at three years of age the first sixteen feet, and I've now discovered that sixteen feet is too much and it influenced the growth, we have to drop back to fourteen or twelve or something," you have complete freedom to say, "I goofed."

That goof is a positive learning experience, whereas in a structured society, if you pruned to sixteen feet and you wrote to your boss that you lost ten percent of growth because I pruned too much, and ten percent over the forty thousand acres we did is "x" volume, you're shot. So you don't allow yourself to dream and probe and do a better job, because you have to protect your job.

But in the kind of environment we had, because no one knew, we had not only the freedom but the direct responsibility to dream. You had to cover the tracks of your dream with real numbers, so that if you dreamed too far or if you didn't dream far enough, or even if you dreamed down the wrong path, you knew to change. These plantations grow so rapidly that when you dream incorrectly, you know within months. I mean, even inventory data would tell you within months.

HKS: It's amazing when you think of the time frame that I was trained as a forester. A twenty-five year rotation. A forester never actually lived with his mistakes because you're long gone before the forest is ready to cut.

CEP: Right. But in this situation, literally, the mistakes are seen on the same timeframe as the mistakes in growing tomatoes.

HKS: There's a certain management cliche that seems to apply to most institutions. The penalties for making a mistake are greater than the rewards for doing a good job, and so people tend to be conservative.

CEP: Absolutely.

HKS: But Ludwig had a situation where he turned that around.

CEP: Absolutely opposite. Ludwig had a saying that as long as you're correct, at least fifty-one percent of the time, we'll make it. His personality was such and his management procedure was such that if you got into a heated discussion and the shoe fell, so to speak, and he said, "Look, it's my idea. It's my money, and we're going to do it this way. I hear you. But we're going to do it this way," you'd say, "Fine."

If he knew that you picked up the shovel and did your absolute best, yourself and all the people that worked for you did the absolute best to make that work anyway, and it failed, he took full responsibility. Never did I see him blame anyone.

On the other hand, if he said, "We're going to move that mountain by six o'clock in the morning," and we only have three guys and only one of them has a shovel, and you didn't try, then it's the same as suicide. To me, there was absolutely no fear of blame and absolutely no fear of making a mistake in the process of learning.

Now, it's incredible that you would have that kind of attitude from a guy who succeeded in getting through the third grade, but in reality he was on a quest for knowledge. He had a lot of ideas that he wanted to prove, and he knew that those ideas could only be proven and become valid with proper information. Yes, he was impatient. He wanted the answer today. But bottom line, he realized that knowledge had to be there for success.

There's really no way to put into words the progress that can be made under freedom. We talked about Weyerhaeuser a while ago. On George Weyerhaeuser's first visit—he had heard about the project, and Weyerhaeuser people had been there before he was there. At the end of the first day, he said that he was absolutely amazed, and he asked me the question, "How could you have conceivably succeeded in doing what we're seeing here in such a short period of time?" I don't remember exactly what I said, but more or less I said, "Freedom." Freedom to proceed based upon best information available that day. And free from committees. Free from corporate structure that requires waiting.

There was another occurrence when we were working with Crown Zellerbach. We were preparing to make our first shiploads of gmelina to Crown Zellerbach. We had done testing on a laboratory scale, but it was now time to do pulp testing on an industrial scale. So we sent a shipload of gmelina to Crown Z in Washington, and it was pulped at Port Angeles and Camas. Once Crown Z had their schedule lined up to where they were prepared to handle that strange shipload, they told us the ship would be there in three weeks.

About the same time we got the message, they sent a guy down to see what stage we're in and get a feeling as to whether it could really be done or not. We took him to the spot where we would load the ship, and it was bare ground. There was no dock. He said, "I have to get to a phone. That ship's leaving. There's no way we're going to load a ship." I said, "See that guy standing right there?" The guy's name is Bob Gilvary. He's a civil engineer from Cornell, and he essentially designed, built all the roads, all the bridges, all the buildings, warehousing, everything at Jari. I said, "Don't worry about it. We'll load the ship." He said, "I have to cancel. It can't be done." I said, "Don't worry about it. Just send the ship."

Well, fortunately, he did that. In three weeks, the ship arrived. He came back. There was a beautiful, permanent dock, still being used today, that we built. He said, "I don't understand it. It would have taken Crown six months to a year to have done that." I had the same response to him that I did to George Weyerhaeuser, that in the presence of freedom and in the absence of fear of being eliminated because you make a minor mistake when knowledge is missing. Now, making a mistake when knowledge is available is a different ball game. That is absolutely amazing what man can do.

Building a dock in three weeks was a normal operating procedure for us, because we were building bridges and roads, and we had built docks before. Small docks, just for receiving supplies and parts and so on. But those things that appeared monumental to what I'll call "normal people" were normal tasks for people at Jari.

HKS: So your engineer wasn't concerned. He knew he could build a dock in three weeks.

CEP: I mean, what is a dock? You go to the woods. You cut some massaranduba, which is a wood that will last forever. We have a pile driver. You haul it to the port, and you start driving piles. You get a bunch of piles in, you put a deck on it. You also drive piles for fenders, to keep the ship from jostling into the dock. I mean, what's the problem? So [chuckling], it's—

HKS: Your engineer was already familiar with the depth of the water and soil conditions and whatever else an engineer needed to know.

CEP: Oh, yes. Knowing that someday there was going to be a plant there, we had already done all of the engineering for the major dock that would be required for the pulpmill. Had done all the soundings, so we knew how deep we had to go to bedrock.

And that was the other beauty of dealing with Ludwig. Even though something was five years away—when we started planting trees, he was not in a mode of waiting five years to start planning a facility, or where it would be. So we were doing port work the same day we were planting the first tree, so to speak.

So engineering work for the facility that would be there, the port facility, everything else, and of course kaolin was built in the early stages of the program, and engineering work had to be done for that. We did engineering work. We did the feasibility studies for a hydroelectric project to provide power for Jari, and that was complete before we ever said a word to anyone on the outside world that it was a possibility.

We had proposals with the World Bank for financing of the hydroproject and the newsprint project, but the basic engineering, to know that it was feasible, was done before we ever brought the subject up with World Bank.

## Amazon Topography

HKS: There were a couple of items you mentioned off tape that I'd like to have put on the record, because it was a new wrinkle to me and part of the stereotype, I guess, of what the "Amazon," in quotes, looks like. About the topography. How adverse it was. It's not flat. It's not swampy. It's rugged.

CEP: Part of the Jari property along the Amazon is flat and swampland. That we used for water buffalo to produce meat for our own population. We also used some of that for development of our rice project. But for those a little bit familiar with geography and have traveled in the U.S., if you visualize western New Mexico and Arizona and Navaho and Hopi country, that land originally, maybe not originally but at one time, was a lake. And the mesa tops that you see driving I-40 west is the original level of the bed of the lake. The rolling land between the mesas is the result of erosion from that original level.

All of the sedimentary region of the Jari property is geologically almost identically the same as what I've just described in New Mexico and Arizona. All you do is apply a hundred inches of rainfall and move it out of the temperate region. We had high plateaus, completely different soils, completely different vegetation. The airport at Jari was built on one of these high plateaus.

In sedimentary soils, it was either plateau or eroded, rolling land covered with forest. Then, when you get to the northern edge of the lake. Originally the Amazon flowed into the Pacific.

HKS: Before the Andes rose.

CEP: Before the Andes rose. The Andes came up and formed a lake. With a hundred inches of rainfall, the lake finally got full enough that it started flowing to the east; hence, the beginning of the Amazon River and the Amazon basin. There are islands that form French Guiana, Suriname, what is now Guyana, Venezuela, Colombia. There was a northern limit to the lake, and so there's a northern limit to sedimentary soils.

These sedimentary soils are deposited over parent bedrock, so if you go to the northern limit of what was the lake and the northern limit of sedimentary soils, or the boundary between sedimentary soils and soils formed in place, and you move south a hundred feet and drill a hole, it's not very deep to bedrock, whereas if you go a hundred miles south, close to the Amazon, where the weight of all of this sediment has depressed the earth's crust, then the sedimentary layer may be thousands of feet deep, down to this original parent material.

You leave north from where we live in Jari and you don't go very far until you leave the sedimentary portion of the Amazon drainage basin, and you get into soils formed from parent rock. From that point all the way north to the border of the countries on the northern portion of South America that I just named. From where you end the sedimentary soils all the way north to where the drainage changes into the Caribbean, it's uphill, and it's rolling highlands into mountains.

The better soils on the Jari property are in the non-sedimentary areas. The soils are much more fertile. There's a lot of variation because all the parent material wasn't exactly the same, but essentially all of the non-sedimentary soils are fertile, and you can grow almost anything. But it's rough terrain.

HKS: So that has all kinds of implications for management. You have the transportation system, harvesting system...

CEP: That's right. In fact, the harvesting procedure in the regions where soils were formed in place is totally different than sedimentary, which is more flat and rolling and well-drained because it's sand. In the region where soils were formed in place, you have a much higher number of distinct streams, creeks, that flow year-round. I mean, it's the same logging thing you're faced with in broken hills in Appalachia. That doesn't fit people's concept of the Amazon.

HKS: What you're describing, your narration here, was this all surprises? Or was it obvious before you even got there that you'd have this variation to cope with?

CEP: Well, it wasn't obvious before we got there. But once I started flying, and we had aircraft and freedom to use them. Initially we did all site selection work on the ground. When we did site selection work looking at soils and taking soil samples, we also did inventory and species distribution. It didn't take us long to find that we could hop in the airplane and fly a region and pick out the better soils in a thirty-minute flight, and know where you're going next year. Now, you still went in and did the groundwork, but we identified from the air, based on vegetation.

Today I can get in an airplane and, based upon the presence of one species, show you where the Indian sites are. I can tell you what soil type it is. Species changes and species distribution are abrupt, as the soils are abrupt. Not necessarily the northern limits of the Jari property, but the northern limits of our activity, our development activity in a way—we did a lot of exploratory work. Close to the northern portion of the property, there's an escarpment. It's an escarpment that basically runs the full distance between the Jari and the Parú rivers.

HKS: It's where the waterfall is.

CEP: No. No. The waterfall is at the break between soils formed in place and sedimentary soils. When you drop out of slowly erodible parent material to erodible sedimentary, that's where the waterfall is.

HKS: I had thought that maybe Club Med would put a resort there. That's a spectacular site.

CEP: Oh, it is. It's fabulous. It was our Sunday afternoon vacation spot.

HKS: I've distracted from you from the escarpment.

CEP: The escarpment stretches from the Jari to the Parú and in places is probably five hundred feet straight down. The soils are different on top on the escarpment compared to the bottom. The variation in habitat, the variation in, quote, "optimum use of land," and the correct species selection—I'm absolutely convinced that it would be ludicrous to think that Jari in three species—gmelina, pine, and eucalyptus—even after twenty-five years, has the correct species on the correct sites. No way. There's too much variation. For proper and optimum management utilization of the resource, they need to continue—and I'm sure they are continuing—to search for the best use, depending upon what soils are there.

HKS: But there's a practical matter. A manufacturing plant has to be able to use this source.

CEP: That's correct.

HKS: So it's not just growing the trees.

CEP: If it were just growing trees, ignoring all economics, it would be much easier. Then the native species would kick in. If it were just strictly for the fun of it. But if it were just strictly for the fun of it, there would never have been gmelina or pine to start with.

HKS: Did you ever think about oriented strand board, as opposed to pulp?

CEP: Yes. Yes. But the finished price is so low that we never could make it work.

HKS: Have we covered the topography adequately, to show the variation and the management requirements, or is there more you would like to say?

CEP: One other thing, just a general statement to give a better feeling for the total region. What I just described was the Jari property. Of course, there are regions that you could go to that it would all be relatively level and nothing but sedimentary. You could go to another region and think the whole world was savanna. Obviously, a region as large as the continental U.S., which if you ignore national boundaries and look at the Amazon drainage basin, it's the same size as the continental U.S. So to imply that the whole place is like Jari would be like my description of a guy getting in a boat and going down a river and then writing a natural history of a continent.

HKS: Right.

CEP: So the Amazon is very varied. The river system is divided up into three categories. You have the main Amazon, which is muddy. Rivers flowing into the Amazon from the south are clear-water rivers, and many of them, the water is just absolutely clear and you could scuba dive and see as well as you can see anywhere. The rivers from the north, like the Jari, are black-water rivers, and there are more tannic acids in the water.

The Jari River looks dark, but if you got a glass of water, the glass of water looks clear. So it's a black-water river, but a bucket of water looks clear.

HKS: But it's not erosion. It's not soil particles.

CEP: No. No. Now, somehow in our environmental mode that the industrialized world is in, if a river is muddy, someone is a culprit and must have caused it. So we have had guests, and we'd be on a portion of the Amazon, and they want to go upstream to get pictures of and determine who is clearing land, creating the muddy water. Well, the Amazon has been muddy ever since the day that it broke out and started flowing into the Atlantic and eroding that part of the world. Somehow we've gotten under the concept that in the absence of disturbance by man, all rivers are absolutely pristine and crystal-clear. Of course, that's very far from reality.

HKS: So the harvesting technology—I don't know how much time you spent on this, but I was surprised, your decision on harvesting technology that you told me about off tape, the high-lead.

## Harvesting Technology

CEP: Well, of course, harvesting is in two phases. There's harvesting of land being cleared, and that wood being used for power. Then there's harvesting of plantations. It's two different procedures.

I'll give this as an example of how our desire to be modern overrides common sense sometimes. In plantation harvesting we proved beyond a shadow of a doubt that the most economic means of harvesting pulpwood-sized plantation trees, whether it be pine or gmelina, is with man and mule. Forget burning carbon fuels. Forget maintenance costs on heavy equipment. Forget the class associated with being modern. The cheapest way was to log and carry. We built racks to go on donkeys' backs—and carry half a dozen sticks or whatever to the roadside.

When we cleared and prepared land for planting in the sedimentary regions, the terrain simply allowed a grid system. We built a plantation maintenance and harvesting road on a grid, so that a tree was never more than a hundred meters from a road. We were super-concerned about equipment being on site, having seen what happened in the initial clearing activity with dozers.

At that stage, we didn't know whether it was removal of the organic, whether it was compaction, a combination of many factors. So we were concerned with damage. Of course, we heard all kinds of horror tales, and we were fearful that disturbance of tropical soils was much more critical than what, in reality, it has turned out to be. We were in the process of devising a harvesting system out of these plantations so that equipment never would need to be anywhere but a road. We could either carry the material to the road or pull it to a road, using lightweight cable systems.

HKS: These were unpaved roads. Unsurfaced.

CEP: Unsurfaced and ungraveled, because of leaching. In most areas without going very deep you could find a layer of laterite. Laterite is simply particles where mineral elements, mainly ferrous elements, have leached out and accumulated and formed a layer. This material is common throughout the tropics, and it makes excellent road surface material. So where and what I'm calling a road in most cases is no more than simply removing stumps and pushing to the side material that didn't burn, so that you had single-lane truck access. Where drainage was excellent, we didn't put lateritic material. The lateritic material was normally used more on slopes, where, through natural erosion, the surface had a higher percent of clay to where you simply couldn't get traction up the hill when it was raining.

HKS: Could you log during rainy season?

CEP: Yes.

HKS: The roads would hold up.

CEP: Yes. Because you would be on a road one or two days. All of the "logging," quote, was designed to be hauled by small bobtail Mercedes trucks that would carry the material, then, to a concentration area, where you would go on larger trucks to the mill or larger trucks on surface spur roads to a railhead. There was never any intent to have all-weather roads throughout the plantations, surfaced for what we think are log trucks.

Because of the grid system and because it wasn't very far to a road, we could cut and haul on a donkey's back to a roadside cheaper than any other procedure. This turned out to be one of the examples among many where economics was overridden by desire to be modern. We never had the first group from a Weyerhaeuser, a Crown Z, Continental, or any other group that we were involved with that could bring themselves to advising going back in time and using man and donkeys.

The normal mentality of a gringo going to the Amazon is that you have to have chain saws. A chain saw breaks down. You have to have a saw shop. You have to have fuel. You have to have spare parts, and they're much more expensive there than here. For the most part, chain saws in the Amazon are a mistake.

HKS: You're cutting small diameter trees.

CEP: Large diameter trees are still a mistake. You can employ three guys to work all day to cut down a tree, and it provides them cash income. Cutting that tree down with three guys is cheaper than doing it with a chain saw. Now, it doesn't fit our image, but it's the best way to do it.

HKS: So there's all kinds of stereotypes. One is that the executive is truly addicted to the bottom-line and if you can show him a deal that saves, he'll do it. But there is an image.

CEP: In reality, that doesn't occur, because we proved—we have the numbers—we did it. We sent shiploads of material, of gmelina, to Washington, to Crown Z. We sent shiploads of material to northern Finland. There were at least three shiploads of material that were produced mainly by ancient procedures, man and donkey. Or just man.

When we were asked the question, "Are you prepared and ready to harvest?", the answer is, "Yes." "How are you going to do it?" "This is how we're going to do it." This was a case where every industrial forestry group—Weyerhaeuser, Crown Z, whoever—every consultant that appeared out of the clear blue had a modern, mechanized scheme for harvesting. And we lost.

This is one of the many cases where we were run over and pulverized and lost. Really, we were right, and we're still right. It is in process of coming to that. Much of the massive harvesting equipment that was there the day the mill started has been sold, not replaced, and they have moved back toward reality. Reality is what I call simple, non-mechanized methods.

HKS: To jump the technology the other way, feller bunchers won't work because the soil won't handle it, the terrain's too steep? I mean, you're harvesting plantations.

CEP: Right.

HKS: In the American South, we use feller bunchers to harvest plantations. We don't use chain saws any more.

CEP: Right.

HKS: But that equipment wouldn't work.

CEP: When you clear and burn, maybe a third of the material is of a species and density that will burn. You have this massive mess that you have to crawl over and under and through. Now, after twenty-five years, there are sites that the amount of material on those sites has decreased to the extent that they now can go in and move that small percentage of what was there originally, and have a clear place where feller bunchers could function.

But because of the economics, the capital cost of the equipment, the damage to permeability of the soils, cost of fuel, cost of maintenance, I still would question, even though they could move freely, I would question the economic feasibility of it.

HKS: Was this before the price of energy went up so quickly in '74? I mean, energy was still relatively cheap.

CEP: No, it was high.

HKS: It was high then.

CEP: In Brazil it's like two-and-a-half times higher than what we're accustomed to. I can show you many examples in the Amazon, and we ended up with one of them at Jari, in spite of bloody battles. That subject is one of the last battles I lost. But I can show you mills that were built by the best engineers, the best technology of the day, selection of the best equipment, setting in the Pacific Northwest would have been a world-class mill that anybody would have been proud of. And it never functioned in the Amazon. We ended up with a massive mill designed for northwest standards, a big log mill that did not have a chance to function from the day it was put on a piece of paper.

HKS: Big logs. So you're talking of native species.

CEP: Native species. You asked a question about oriented strand board. All of the inventory that's there, probably fifty percent of total volume has a density in the range of 1.0, and to kind of get you worrying—hickory is probably 0.6—so there are many species and a lot of volume that will not float.

There are many species that have high silica content that, under today's technology, are simply not usable. You saw them, it ruins the blades. You chip them, it ruins the knives.

HKS: Never heard of that one before.

CEP: If you look at World Bank inventory of the Amazon, people say that it's the world's wood basket. Well, that is false, and not even anywhere close to reality. A lot of the inventory work that was done was based on total volume. It's an over-mature forest, so you measure a large tree and it has tremendous volume in it, but you cut it down and it turns into powder. Because you have an outer ring.

You're familiar with over-mature stands in the Pacific Northwest. I have harvested stands, old-growth stands, that had as low as thirty-seven percent usable material. It was just simply too old to be used. Of this tremendous wood basket that is in inventory numbers, a high percentage of it is over-mature and the trees simply not usable, regardless of what species it is.

HKS: Can't get it out? Or has silica in it? Or—

CEP: No, it's too old. It's like an over-ripe watermelon. It's hollow. When you cut it, it shatters. It's just simply over-mature and non-usable, under any form.

Then there's a percentage of trees that are species that do not have a cylindrical trunk, trees that we call "crazy wood." The surface is convoluted. There are many common names for them, but they make up a significant portion of the total stand.

There are trees with such high silica content that under today's technology they're non-usable. There are trees without the silica but such high density that they can't be used in that under today's trades, if you give a guy a two by four that weighs three times as much as any two by four he ever picked up, and he can't use a gun to nail it, then of what use is it? Yes, it's beautiful. Yes, we can harvest it. Yes, we can figure a way to saw it and dry it and surface it. But if you can't nail it, and it takes two guys to pick it up, who is going to use it?

We had small mills, Mighty Mite portable mills, that we produced the wood for our own use. We set those mills up in every forest type that we had, and when total inventory was a tremendous volume per acre, we never exceeded three to four thousand board feet, and the average was more like eighteen hundred board feet to the acre, usable.

HKS: Eighteen hundred.

CEP: Eighteen hundred.

HKS: That's not very much.

CEP: It's nothing. And in fact, in big trees, there is no economic means of building roads and logging to recover eighteen hundred board feet per acre.

HKS: I can understand that.

CEP: The concept is that there is tremendous volume there which, in reality, doesn't exist. It's like there's tremendous volume in the Pacific Northwest today. But it doesn't exist as far as use is concerned because it is out of production. There is volume in the Amazon, not what people think it is, and it's basically non-usable.

HKS: It explains the numbers which show something like two to four percent of total tree removal is for commercial wood production. The rest is for land-clearance, for agriculture.

CEP: That's correct.

## Amazon Belongs to Brazil

CEP: You have to ask yourself the question, Asia developed in wood use. Africa developed in wood use. With all these trees. You can get in a 747 and fly six hours and see nothing but trees. Why hasn't it been developed? With minor exceptions, there is no major forest products company there, and never has been.

One additional problem is that of that small percent per acre that is usable, it's distributed among hundreds and hundreds of species. So how do you develop an economic use for that natural resource under those limitations?

HKS: Basically what we know so far is that you need a Jari type project, where you clear the land and then put in a plantation. Otherwise, forestry is not a practicable enterprise in the Amazon.

CEP: Yes, there are regions where mahogany as a species is predominant. Yes, there are regions where another species is dominant. But for the most part, the diversity is so great that utilization.... There should be no fear of anyone going in and clear-cutting the Amazon for industrial utilization of the wood. The resource does not lend itself to that.

Clearing is a result of population pressures. It's not pressure for wood. I mean, 99.999 percent of every tree that gets cut gets burned where it falls, because there is no other use for it. The reason clear-cutting occurs is population pressures. There are no major land-clearing projects remaining. World Bank isn't financing any. Brazilian Incentive Programs is not financing any. So what is the clearing? It's population pressure.

The only way to control the clearing that's occurring is to develop products that come from that resource so that it doesn't have to be cleared, and in so doing, provide jobs so that people can eat without having to clear.

No amount of wailing from the thirty-fourth floor of some building in Chicago, no amount of anything is going to change the reality of that land being cleared. So if we as an industrialized people can get together and quit wailing about something that happened thirty years ago and start a positive approach to develop those lands that were clear-cut and abandoned, into a job-creating asset.

They should stop efforts of banning the use of any product from the region, because I see no difference between the southeast U.S. and Brazil or Lake States region and Brazil. When the South was first populated by pioneers, there was no market. They had to have something to eat. So they cleared and burned. Yes, they used some logs for a house and a barn and a fence, but there was no market. That material that we look at today and call a resource was an impediment and a barrier to his survival. There simply is no difference. That's what the Amazon is facing today.

HKS: Is there, by your observation, with individual families clearing their three acres, is there enough of that activity to generate the fear and concern you read in the popular press? Or are there large agricultural companies—

CEP: There is, and yes, there are large companies. You get in the State of Acre, for example, which is western side, you're out of the sedimentary region into soils like I described in the northern part of Jari, very fertile. The chance that those fertile soils can be allocated a value of zero and that those lands will not be cleared. The chance of that occurring, at least in my mind, is zero.

That would be equivalent to asking in the early days of the development of the U.S., asking the 1880 population of the U.S. not to clear another acre in Indiana, Illinois, Ohio, and that the U.S. should not have a corn belt.

I don't like to see it. But we as a nation cannot ignore the reality of people, masses of starving people, when we look at policy related to the Amazon. Number one, as I mentioned earlier, the Amazon is not ours, and it never will be. Brazil as a nation is interested in developing its resources to feed its people. There is no way that any amount of wailing is going to change that.

## Brazilian Forest Practice Laws

HKS: That leads into another element of the same thing. Brazilian forest practice laws—I'll call them that. I'm not sure what you would call them. How does that influence the way Jari operates?

CEP: The way Jari operates probably the influence is nil, because any acre that Jari clear-cuts today, Jari is not in an expansion mode. They have all the acres they need to feed the pulpmill. The only acres being cut and planted are those required to feed the power plant, and that is the reason that the hydroproject has been approved by the government, so that land-clearing for power does not have to occur in perpetuity. There has to be an end.

Of course, a power facility wears out. Just because a power facility is depreciated out and gone doesn't mean that the plantations have to stop. So they're on the path of replacing power from burning wood to hydro. Any acre that's cleared is replanted. There's really no activity there that requires observance of any law. I mean, the law says if you cut trees, you plant. Well, obviously, they're doing that.

HKS: From time to time, a Brazilian forester would come out and do oversight procedure.

CEP: Right.

HKS: That you're doing what you're supposed to.

CEP: Right. Now, the law requires that if you cut a tree you plant, I don't remember, three or five to take its place. That obviously doesn't apply in a plantation, because when you cut a plantation tree, you plant one to take its place.

What the law has allowed, and this is what GP did when they owned property there—they no longer do—but there were forestry companies established. Let's say GP cut "x" number of trees. They simply paid a forestry company to go out and plant their quota of trees, and they did not necessarily own the new trees planted. So for a guy that was conscientious and had a long-term outlook, this was a bonanza, because big non-Brazilian companies that had to follow the law had to pay cash. So he could put in a plantation that didn't cost him anything.

They simply do not have the money to police and be sure that it's carried out.

HKS: So you bought the property. You didn't have to file a management plan and pay a fee or something like we would in the States.

CEP: No. No. There are ample laws on the books to cover anything that's happening and anything that might ever happen. The problem is, there is no money for governmental agencies, and so in the Amazon there are people responsible for enforcing forestry regulations, but they have no budget. Not even a budget to pay their salary. Not even a budget to buy the gasoline for the boat. Not even a budget to buy the boat that they're supposed to ride in.

The only source of income that that organization has is fines. Basically, if you are doing anything down to being an individual in a canoe with three boards, you will be fined. You don't have to be guilty of anything. But in order for that Brazilian organization to survive, you will be fined.

HKS: Is it the same in southern Brazil, that they don't have a budget?

CEP: In much of southern Brazil you couldn't tell the difference between there and being around Savannah, Georgia. There's an open, honest, economic incentive to be doing what's right. You're already making a living off your plantations, and you need more. So yes, there are people there, but it would be like policing Oregon forestry laws.

In Oregon, if you're going to cut, whoever that's responsible has to put up a bond that is economic proof that you have the capability of replanting. If you don't replant within a given amount of time, the state replants and calls the bond. It doesn't take many people and much of a structure to see that planting is done. I'm sure there are instances, but I don't know of any instance where a bond has been drawn on.

Policing to that extent, where it's economically sound and feasible to fulfill the letter of the law, is not a major problem. In the South it's developed to the extent that it's either land in agricultural production or it's in plantations.

## Recruiting Workers to Work in Isolation

HKS: When you interviewed people for jobs, how did you characterize the job? You looked for some kind of response. The kind of questions they asked would tip you off what their concerns—

CEP: Job interviewing was the most inadequate feeling of any and all the responsibilities I had. That was the most inadequate, because they have such a preconceived idea that they hear your words, and they can write them down, but it does not register.

To give you an example, I hired a young geneticist who was one of my graduate students. I left when he completed his degree. I, very plain, in the English language, explained to them what the conditions were. At that time, which was early stages, the supermarket was a shed, no sides, and in the supermarket were bags of beans, bags of potatoes, cans of powdered milk, basic items. Rice. And that was the beginning and the end of it. And you got fruits in the forest. We hauled in daily, on our airplane, onions, tomatoes, fresh vegetables. You got fish from the river. But the supermarket was as I described. The housing for them was a house about twice the size of this room, and screened in.

They arrived. I took them to the supermarket. She stood there and cried. I took them to their house, and she cried uncontrollably. And after she got through crying, she said, "I'm sorry. I heard the words that you said. You told us exactly what the supermarket was, and that's what it is. You told me exactly what the house is, and I see that's what it is. But somehow, in my language and my experience, I could not relate to it, and it did not register. Now that the reality of it is before my eyes, I'm in tears."

That's a vivid example. That was the truth and the situation. Whether it's professionally, and the guy's job you were explaining to them. The fact that you were telling the guy that he has to learn a language. He's never spoken another language before in his life. He's never studied languages. You're telling him he will learn a language or be gone. He's accepting that and saying, "Yes, I will learn a language." It's a new experience.

The thing that got many people was the cultural shock. You can explain cultural shock to people, and it's meaningless. We had people pack up and go to the airport because we were not taking care of the local population. The local population had health needs. The Jari health budget, our hospital and our health budget, was greater than the Brazilian federal government spent on the total Territory of Amapá. We, in our isolated situation, spent more money on people's health, free of charge, than the federal government spent on the total territory.

But there are still limits upon what you can do. If a sick woman showed up in a canoe at our hospital with a dying baby, it wasn't our responsibility, but guess what. We took care of that baby. But still, within that social framework, there are limitations. We could not physically take care of the total northern Amazon region. We couldn't take care of Brazil. We took care of those things that confronted us, but still there were tremendous needs. If you're not taking care of those total needs, you're the bad guy. So we turned into the bad guy, and this employee will no longer work for bad people, and he gets on an airplane and goes home.

So you have all kinds of problems associated with the society around you.

HKS: Did the supermarket become more sophisticated and better stocked in later years? And housing improved?

CEP: The supermarket today at Jari, you'd think you were in United in Brazil. Anything you can buy is there. But it takes time to get there. It takes demand. You can't have that kind of supermarket when you have fifty people.

HKS: The locals, the workers, that was the diet, rice and beans and flour, that's all they ate anyway.

CEP: That's all they knew. Yes. It's all we ate anyway. [chuckling]

Very rapidly, we established the basics of a nice place to live. From a standpoint of medical facilities, water, sewage, roads, supermarket, communications, today essentially are there. And has been for a long time.

But the limitation for people being there is isolation, not things. At first, we had almost nothing, a cleared spot on the edge of the riverbank and a supermarket, a shed, and a lot of times not even a doctor on the site. On my oldest son I can show you four scars—he was a kid that was afraid of nothing, so he was always getting beat up and hurt—where his daddy sewed him back together, and his mother was unhappy with the sewing job. Those would always happen at times when there was no doctor there.

Even when physical conditions were bad, compared to when physical conditions were good—with swimming pool and movies and everything—I could detect no difference in the attractiveness of the place or ability to hold people. Isolation was the problem, and isolation is the same whether you're living in a shack or a mansion. It may take a few extra days for isolation to get a hold of you in prime surroundings, but if you're susceptible to it, it still gets you.

This gets back to the thing I said about if you have to have your hair fixed every week, it won't work. So you're there. You have freedom of movement. On Sundays it was a going thing for almost anybody in management that had access to a vehicle to take their wife and kids and go to the end of the road. Because we were always building roads, you could go a kilometer further this Sunday than you went last Sunday. But still, you couldn't get out.

The road might be a kilometer longer, but it was still jungle at the end. The feeling and the concept of isolation affects some people much greater than others. Even with prime world-class physical conditions, some people can't survive.

HKS: There's no way to know in advance.

CEP: There's no way of knowing in advance.

HKS: If you never had that experience.

CEP: Right. To me, I got on the airplane and went to Belém when I actually had to. If somebody took a switch and ran me off, I went to Belém. But I would much rather be in the jungle than in Belém, but most people are the opposite.

It depended a lot on husbands, also, in that if the husband could not realize what his wife was faced with, they didn't survive. If he realized that, while he was out beating the bushes and learning and seeing all kinds of things and having all this tremendous challenge, she was staring at the same four walls. If he couldn't grasp the significance of that, they didn't survive.

HKS: Much like the early settlers in Oklahoma. The husband was farming; the wife was in the house. No neighbors to talk to.

CEP: Right. It's exactly the same. On occasion, I would see stress developing. Understand that men will get so tied up and so challenged that nothing else exists. I mean, they don't write their momma back home. They come in at night dead-tired, and their wife is there, and they say, "Hi" and take a shower and go to bed, and get up at four the next morning. They know they're married, but she doesn't exist.

Yes, they love their wife. But they never dreamed in the wildest dream they ever had that they would ever be able to participate in something like this and have such wide-ranging responsibilities and be able to allow their professional mind to expand at its rate of capacity rather than at an organizational rate, or be pigeon-holed.

This was one of the responsibilities of my wife. When she would detect those kinds of things developing, she would advise me. It normally did me no good whatsoever to tell an employee that he needed to take a few days off. You might get a response, "Sure," on his way out the door to go have his fun. On many occasions I would arrange something for them and tell the wife, and the guy comes home one night and she's so excited because he's taking her somewhere, and he doesn't know anything about it. But then it hits him that something has happened, and she's all excited and stirred up, and there's no escape. He has to go. [laughing]

HKS: Is it part of the compensation package a plane ticket to the States once a year? Or some such thing?

CEP: You had a month's vacation, if you were non-Brazilian. You had a month's vacation in the States every two years. If you were Brazilian, you had a month a year.

Inside Brazil, we were always limited on space on the airplanes. You'd go on a boat any time you wanted to. But you normally could go to Belém once a month for a weekend. You'd leave Friday afternoon and come back Monday morning. But some people, including myself, didn't use the once a month to Belém and didn't want to. Some people needed it every other day, which means that they didn't survive.

HKS: A few Australians I know have a sense of isolation. Even though they can go places and everything is there you could possibly want, it's just that they are so far away from England or whatever they identify with, it's a problem.

CEP: Yes. I think anything you could ever want is in Brazil. Somewhere. But for most people, it's two thousand miles away.

The development of something like this confronts and creates tremendous social problems. To get clearing and plantation work done, we would bring in ten thousand employees at a time, that have to be clothed, housed, taken care of medically. We built plantation villages, and this is a whole other subject. But we built plantation villages, each village with its own health facility, school, recreational facilities, supermarket, everything.

HKS: Is that an example of what you're talking about?

CEP: This [showing a photograph] is Monte Dourado—the main town.

HKS: That's more advanced or more plush than what we were talking about?

CEP: Right. Plantation villages I think came along after this was published. But just the sheer size of the thing, you could transport workers from where the main town was to the furthest plantation. I mean, you'd get there at noon. Eat your lunch, and be able to come back home. So we built satellite or plantation villages out in the plantations, and moved in people from northeastern Brazil. These people came from a region extremely dry. It would be like moving a family from Arizona.

They had never had a permanent roof over their head, never seen running water, never seen ice, never seen a doctor. No child had ever been to school, and you move those into a planned community, with school, with water, with essentially everything that we in our society would expect. It is a tremendous traumatic experience for these people.

We designed the place so that each house had a garden spot. We had an agronomist. When a family would show up, each family would receive banana plants, manioc plants, fruit trees. The agronomist would work with these people, training them how to grow their own food.

Each village had an agronomist, had a trained social worker, a girl—in each case it was a girl with a degree in sociology—to help these people adjust to a new form of life.

HKS: Why was the labor recruited from that particular area? Is that the closest population?

CEP: Yes. See, Jari was established in a region. Yes, there were people, but they were gatherers, and population density extremely low. You'd see a scattered house along the river. The Amazon, except for specific cities like Belém and Santarém and Manaus and Iquitos and so on, except for specific cities on the river system, the Amazon is devoid of people.

Jari was established in a void region. Yes, someone had walked over the land, gathering Brazil nuts before. And yes, somebody had been up the river in a canoe. But basically it was devoid of people. So the masses of people required for Jari to function had to come in from somewhere else, and the logical place. You don't even have to go back in history. The northeastern part of Brazil is extremely dry. Some years it rains, and some years it doesn't. When it rains, it floods; and when it doesn't rain, there's just no water.

HKS: So that's the area north of Jari you're talking about.

CEP: No. It's east of Jari. Northeast. It was the Territory of Amapá, east of Jari. When it doesn't rain, people put a stick and a bag over their shoulder and head to a city. You have mass migrations starving people into places as far away as São Paulo. But that's still home. When it rains, they head back home.

It's a region that is way overpopulated in terms of what the region can support. We need workers and permanent employees, and they need permanent home and income. We transported people by the tens of thousands from northeastern Brazil. We had hiring stations throughout that region. Employees that went out and set up shop and they'd hire people.

So there was a steady flow of buses, boats, moving people to Jari and people out. In the early years, we hired tens of thousands, and they were temporary. Once we built the villages, to bring your family and come to Jari, you had to have worked at Jari in previous seasons. You couldn't just think, "I think I'd like to live in the jungle." They'd never seen a jungle before. If somebody came and wanted to move their family and had not worked there before, they didn't qualify. The only people that moved were those that had worked there as seasonal workers one, two or five seasons, and decided it would be a good home place.

HKS: Reading through the articles, and I realize you had to pick a particular year, maybe the time of year, but what was the population of the area? I mean, I've read ten thousand. I've read thirty thousand.

CEP: Originally?

HKS: No. When Jari was up to speed.

CEP: Population is a hard thing to get a hold of. Going back to just the subject of hospital. If you live on a river fifty miles away and you hear of a place on the river where there's a hospital, they give you an antibiotic if you walk in with a dying child, they save the child, what are you going to do? You're going to get in your canoe, and you're going to move.

There's a city that developed across the river from Jari that we tried to prevent and could have prevented. That city has more people in it than the total population of Jari and its employees. If you go there, what you see is the attractive city that lives off Jari. It's where Jari employees may spend their money. Get off work and go across the river and have a beer.

Regular commerce has developed. If you're a Jari employee and you want a refrigerator, in all likelihood you go across the river now and buy it. It started just as covered shacks along the river. So the population of that attracted city, at both Monte Dourado, which is the main town, and the port of Munguba, that population is greater than the Jari population.

If you count that population, you're probably looking at fifty thousand people that obtain their livelihood directly from Jari. Jari does not have that many employees.

HKS: In that city across the river, whatever benefits a Brazilian citizen gets theoretically is available there, but practicality—

CEP: It's all in theory. In fact, if you live in Belém it's in theory.

HKS: Is that right?

CEP: Yes. If you're a Brazilian citizen, you pay into a social security system which includes medical. If you get sick, in all likelihood you stand in line for days to get to the front of the line for them to tell you they can't help you. So it's a horrendous situation.

## Agroforestry

HKS: Let's turn to agroforestry.

CEP: Okay. In much of the developing countries, specifically in the tropics, agroforestry is and has been a subject that FAO, USAID, and various organizations have pursued in attempts to look at food and fuel production.

At Jari, we started planting pine in '73, and pine was allocated to the sandy loam soils. One of the problems in the tropics is high rainfall and consequently leaching of nutrients from soils that tend to be highly permeable. Of course Jari was in an isolated region, so one of our objectives was to grow the meat and vegetables and as much food as we could for our own population.

We had water buffalo in the *várzea*, or floodplain, and we also had a cattle program. In the gmelina plantations, once an area was cleared, burned, and planted, within a year on good soils you absolutely had complete cover. On sandy soils, it was probably eighteen months before you had coverage.

Pine starts much slower. When you plant pine, it sits there for a period of time, establishing a root system, and then growth begins. But even when it begins, it grows slower than gmelina. So at the end of the first year in a pine plantation, you probably don't have any more than maybe ten percent coverage of the soil. We had airplanes because of our other agricultural projects. As soon as an area was cleared and burned, we would sow grass seed. We used species that are bunch grasses, and bunch grasses, rather than having feeder roots on the surface, are deep-rooted. So they can function as nutrient pumps. We would sow the grass about the same time as we would plant the pine. But the grass, during the first season, would also be coming established and not present significant competition to the trees.

The second season, pine jumps out of the ground, so to speak, and grows rapidly and gets high enough that the bunch grasses, even though they will be more than six feet tall, the pine would never be in a position to where it would not get direct sunlight from above. It might have some side competition from the grass, but it would never have competition for sun from above, so we didn't feel that we were negatively influencing the growth of the pine.

In this process, this small pine tree could not reach out and grab all of the nutrients that were released as a result of the fire. But immediately planting grass, the grass could grab these nutrients, and the portion that was leached, the deep-rooted grasses would get the nutrients and pump them back up to the surface.

So the second season, then, the pine is tall enough—the cattle would not damage the pine—and the grass is established, and we would rotation-graze the pine plantations. We would fence it in blocks, and you run in a large number of cattle, graze the grass to feed the cattle and minimize the competition to the pine, take the mass of cattle off, put them on another area, the grass comes back, and when the grass gets to the point that it needs to be grazed again, then you run cattle in. Then each successive year, as the pine grew and shaded out the grass, then there was less and less cattle, to the point that by the fifth year, depending on site, there was not enough grass remaining, because of shade, to justify opening the gate. And then you take the fence down, and you no longer need the fence until you clear-cut and plant pine again.

In this manner, you help the biological community get re-established after a fire, because you're providing shade and humidity at ground level much sooner than you otherwise would. You essentially prevent leaching. In fact, I think what happens is the deep-rooted grasses, as a nutrient pump, is one of the reasons that growth is better the second rotation than the first.

This gets back to a statement made previously, that all of the horror stories that I know of related to tropical soils can be true, in the absence of logical management. But with proper management, the horror stories are not valid, and they simply disappear.

HKS: Did you ever run into a problem where you had more cattle than grass? You had to import hay or something?

CEP: No.

HKS: Because trees were your number one priority.

CEP: Right. No. Because with forty thousand mouths to feed, we never had enough meat. So if there wasn't enough grass, there were plenty of hungry people. So we never had the problem.

HKS: It was growing your own food. There was never any really planned extra meat. This was part of one of the by-products of Jari.

CEP: No.

HKS: Or rice.

CEP: Well, rice, yes. And here again—

HKS: Did you ever think about—

CEP: The meat is not exportable. Yes, you could put it on a boat. But no, American people accustomed to the quality meat that we have here would not masticate enough times to get it down. [laughing] Yes, it physically could be put on a boat. But in general, not marketable.

HKS: I'm sure somewhere in forestry school you read about the controversies of allowing cattle to run free or not on forest ranges, what the damage it did to seedlings. So you must have had some apprehension when you first tried this.

CEP: No, because we had control.

HKS: The grass you knew was preferable for the cows that the seedlings, they wouldn't nibble off the terminal buds.

CEP: Right. The grass is absolutely luxuriant and high-protein, and pine is not. In our case, there were plenty of stumps and trees that did not burn for every head to have its own rubbing post. [chuckling] So the physical damage was not a problem. The other concern, going back to our school days, was compaction. The soils that we're dealing with here have a high enough sand content that they're essentially non-compactible.

Now, if you ran large numbers of cattle continuously, under those conditions you in all likelihood would alter and damage the structure, and in altering the structure, affect the biological community. But we ran such high numbers of cattle for such a short period of time that we didn't have those kind of difficulties.

## Wildlife Concerns

HKS: Was wildlife damage ever an issue? I don't know what animals live in the jungle that were significant. I mean, in the States you have elk and deer and all that.

CEP: You're familiar with virgin stands in the Pacific Northwest. They're almost devoid of wildlife. It's a closed system. There's very little sun on the ground. There is very little food. In people's minds, this pristine, untouched environment must be loaded with wildlife. It's not. Wildlife is rare.

We keep hearing this word "diversity," and it's used from every direction for every purpose imaginable. But I have been on site-selection trips where we just take a compass and head out in "x" direction for days and go a hundred kilometers, and cross rivers, swamps, over mountains, just whatever is there in that straight line, because we're trying to determine what is there.

In let's say a fifty-kilometer trip, one might see three troupes of monkeys, and none of the three the same species. The next week, you might strike out on another line in another direction and see one, two or three troupes of monkeys, and none of those three species be any of the three that you saw the previous week. So the number of species is just absolutely startling.

Of course, things that people think about in the jungle, for example, is snakes. In the Jari region, snakes were so horribly bad that I was there thirteen months, in the woods almost every day, including Sundays, and Sundays with the family roaming around learning, was there thirteen months before I saw my first snake. Every strange, unusual snake that you've ever seen in the *National Geographic* magazine exists. It's real. But there are so few in number that it's unusual to see the same one very often. Once I was there for years, and I knew a little more about snakes and understood a little bit more about the habitat and where each lived, then, by knowing and searching, I could see more snakes. But as a casual observer, the only snake that you would ever see might be one sunning in the road.

HKS: You didn't have bird damage to your nurseries. Wildlife damage was not an issue.

CEP: No. In the forest, there's jaguar, there's what we would call here "black panther," there's ocelot, there's tapir, anteaters, deer, there's several kinds of the world's largest rodents, the cotia or agouti, the capivara, paca—most of them good meat. The population there eats essentially anything that moves.

We always had concerns and criticisms based on clearing and destroying wildlife. Our world is much more versatile than people would want to think it is. You can have a horrible disaster, and the environment recovers. Not necessarily in exactly the same way, but in our plantations, whether it be gmelina or pine, because of the edge effect of roads, so that wildlife have a place to be and the forbs and fruit-bearing plants that would develop wherever light occurs along roads. As an example, the deer population just explodes.

HKS: There is deer there.

CEP: Yes. Wildlife comes back into these plantations with a roar, so to speak. Again, not in the same percentage and the same species distribution, but the numbers, if you counted numbers of individuals, including snakes, the population is several times greater in a plantation a few years old than in a native forest.

Birds that one has learned are attracted to and normally can survive in the upper canopy of the natural rainforest are happy dwellers in the plantations. We don't know enough about it. They may come there and nest there because their natural enemies are not there. They go back and forage in the surrounding rainforest. We don't understand enough of those combinations and inter-relationships and things that happen. But the wildlife is certainly there in large numbers.

HKS: My only experience was a one-day tour of the La Selva operation of OTS, Costa Rica. You could hear the birds calling. It is very exotic and all that.

CEP: The outside concept is that the whole Jari world, so to speak, was cleared and planted to a monoculture. Less than ten percent of the ownership is in plantations. There are areas set aside for Brazil nut reserves. The timber along streams in many areas is left. And, of course, in a plantation where the objective is an economic return, you don't just start at one corner of the property and start clearing and go to the other. You plant those areas that give you the greatest chance of growth for the particular species that you're planting, and at minimum logging costs.

So when you have a region as diverse as Jari, there are a lot of strips and a lot of large blocks just not touched. That does not mean that there are not large blocks of plantations, because there are. But it's certainly not a situation where you start at one corner and clear everything.

HKS: You read in what I'll call the "concerned" literature about the need to maintain corridors, that animals need to migrate for various reasons in their life history. It's the elimination of corridors that's more devastating to the wildlife than the clear-cuts and so forth, because they need to move around. They wind up in islands, and they can't move to their summer sites or winter sites. Was that part of your management plan?

CEP: You're dealing with a very large region. Yes, there are corridors, but when one thinks of clear-cut and the number of acres that were cleared at Jari, in your mind you see that many acres cleared. Let's make it easy and say three hundred thousand acres are cleared and you're on a ten-year rotation, then only thirty thousand acres is cleared at any one time. We never observed that in, let's say, plantations after the fourth or fifth year, wildlife didn't hesitate in coming in, living, going through.

I certainly don't understand the edge effect, but one of the best places to observe the macaw, which is a large parrot, is on the boundary roads between plantations and the untouched rainforest. Part of that is that simply the rainforest is closed and so high that when you're in the forest it's rare to see one of them. You can be going through and make noise and disturb them, and they leave and make a noise, and you recognize the sound, but you never see the bird.

HKS: Is it worthwhile for us to talk about plans for expansion of the second mill and hydropower? Did that affect the decisions that you were making because you were in phase one, in your mind, or not?

CEP: Well, [pauses] my mind is still clicking on—

HKS: Agroforestry.

CEP: You're bringing up a broader subject that we need to discuss. That subject we started earlier. That subject is related to why we left and why the property was sold.

HKS: Let's talk about that.

## Plans for Expansion and the Sale of Jari

CEP: We had planned a forestry expansion and planned a fifteen hundred ton per day newsprint facility and a facility to reduce bauxite to the alumina stage, because we had, as I mentioned earlier, developed a bauxite mine on the Trombetas River. We were going to barge the bauxite to Jari, because Jari would have a source of power.

The aluminum processing facility, the hydroproject, the fifteen hundred ton a day newsprint plant was essentially all in one presentation and in one package with the World Bank. From the beginning, the concept was that Ludwig would put in his money to prove that the program was viable. He would determine what the resources were in the region, which we did. We would prove that growing wood on an industrial scale could be done. We'd build a town, hospital, schools, everything. But once all that was proven, then it was time for government to kick in.

So we did all of the planning and engineering. We had the dam site cored, everything to the stage that we had halfway decent estimates of costs. One of the keys in the total equation was the government kicking in and taking up the responsibility of the infrastructure, so that we as a company would no longer be responsible for being mother and father and everything to everybody all the time. I never did feel and Mr. Ludwig never did feel that it was bad intent on the part of anybody in government, but worldwide occurrences...

Remember at this stage Brazil had gone through and was still in the midst of the horrible economic beating they had taken as a result of the international oil embargo. They had borrowed billions of dollars from the industrialized world for economic development, and for the most part, compared to many other countries, in Brazil you can see where the money went. I mean, it went into airports and education and infrastructure. We didn't get the money back, but you can see where it was applied, at least.

So they were hit with extremely high oil prices because they were importing ninety-two percent of their requirements. And double-digit interest rates. They were literally on their backs. The compounding effect of those two factors were at their maximum when we were ready to proceed in the second stage of Jari, and the second stage was over a billion dollars. We had meetings with government, the president, various ministers. There was not a single one of them that did not feel their responsibility and there were none that did not want to proceed. But they simply had zero money available and could not see how they could politically survive by applying that much capital in a developmental program when the population in general was going backwards.

It became obvious that if we proceeded we would have to proceed on our own, as we had done from the beginning. The decision was made that it was not logical to do that. Once that was decided, then we obviously dropped negotiations with World Bank and with everybody else, and embarked on a path of selling the property.

HKS: Mr. Ludwig wouldn't approve of saying "What if?" If there hadn't been the oil embargo, it looked feasible. The whole thing could have worked. Brazil would have stepped in according to plan. World Bank would have helped, and so on and so forth.

CEP: Right.

HKS: It was a combination of factors. It wasn't, as one of the authors wrote, "jungle madness" to have been there.

CEP: No. Not at all. There is a Ludwig counterpart in Brazil that I've mentioned. His name was Antunes. He has many holdings, but he is Cummings Diesel-Brazil. He is what used to be Hanna Mining-Brazil. He is Swift Foods-Brazil. He is Scott Paper-Brazil. A long list. He's a major industrialist, and when non-Brazilian companies wanted to come into Brazil and set up operations, he would go in with them. They would bring in their capital. He would own part of the new venture. So all those companies that I named, plus many more, he owned, controlled, depending upon circumstance.

The large-scale pine plantings that we started in the Territory of Amapá belonged to him. He, being Scott Paper-Brazil, being in the wood products business was not foreign to him. He and Ludwig were close friends. His organization took fifty-one percent of Jari and a consortium of twenty banks, the other forty-nine percent.

The media would lead you to believe all kinds of erroneous things. Remember that I said previously that Ludwig would put in whatever amount of money of his own that was necessary to get a project off the ground, but the minute that it was feasible, then he would go the normal financing route. So, just in global numbers, Jari cost somewhere in the range of a billion dollars.

The pulpmill was built in Japan at shipyards that Ludwig controlled immediately after World War II, and it was the shipyards where he built the bulk of his fleet. He knew the people. He knew the management. Knew the capabilities. The pulpmill was built at a time that Japan was in tough economic straits, and so we negotiated a Japanese export-import bank loan, because they needed to create jobs, shipyards sitting there idle. So we negotiated financing for the pulpmill, power plant, chemical plants, the whole deal, at low interest rates and zero interest until the mill was running. So interest didn't even start when it was completed in Japan, or when it was towed across the Pacific. Interest started once it was running.

The Japanese export-import bank loan then was guaranteed by the Brazilian National Development Bank. Of course, what I'm describing here is not all of the financing that was done, but it's an example. When we sold Jari, it was a transfer of financial obligations from us to the new organization that was accepting that responsibility. If you go to *Fortune* magazine or somewhere, whatever the total amount of the pulpmill or whatever it costs—say six hundred million dollars—Ludwig lost six hundred million dollars.

If you accumulate all of the so-called losses, his losses were tremendous and put him on his knees. Really, all it is financially is that if you have an asset that you owe a dollar on, and that asset has a value of a dollar, then you have a dollar on each side of your balance sheet. If someone else assumes the liability of that dollar, you no longer have that dollar on either side of your balance sheet.

HKS: Would it be proper for you, or do you have the knowledge of coming up with the real number, in the ballpark, of the loss or the gain or was it a wash?

CEP: Yes, I do. But I won't.

HKS: Okay.

CEP: Ludwig is no longer here. He passed away last year. But I wouldn't tell. I wouldn't want to do that then, nor now.

HKS: It was a billion dollar investment. You're definite about that.

CEP: Let's put it this way: About a billion dollars went into Jari. The pulpmill was considerably more than half, and that was a financing package. There were other financing packages that were also assumed in the transfer of ownership. There were all kinds of financial arrangements.

For example, one of the largest iron ore export projects in Brazil was a company called MBR, and that was jointly developed by Antunes and Ludwig. I guess the only thing I should say is that at the time that Jari was transferred, there were other transfers of ownership in different companies and organizations, to where no one except probably Ludwig knew the end result.

This is speculation, not fact. But let's say he owned "x" percent of MBR. After the close of the Jari transaction, he owned "x plus." So if you look at Jari in a vacuum, yes, he lost money. But not anywhere—because of the various financing packages and their assumption by the new owners—no, he didn't lose anywhere near what the media would think. And because of various transactions not directly connected to Jari, his losses were even less.

HKS: I'm thinking as an editor here. Somewhere down the road, this interview may be combined with others for publication. One of the issues that needs to be addressed are the accusations in the popular media we've been talking about. You have dealt with a large number of those.

At some point, we're going to have to come back to an estimate of some dollar amount. It is still an incomplete story, because so much has been made out of the bad judgment that went into it. You've talked about the oil embargo had more to do with it than anything else. And all the rest of it.

CEP: I contend that there was very little bad judgment. When you look at it from a standpoint of a banker lending other people's money, then perhaps you'd say it's bad judgment, because we had all the facts available that day, but it was still no facts. You had to learn as you went. I do things today based on a hunch.

I have a small company that occurred as a result of a conversation one day with my local banker. I did no planning. I did no budgeting. I did absolutely none of the normal things that one would do to establish and determine if you should embark on this new endeavor. That is the only company I've ever owned that has been profitable from that first day to this, and it's been fifteen years. I followed zero of normal business practices in doing that.

Let me say it this way: The bulk of the criticisms against Ludwig is that he was an individual, who expressed himself. And you're not supposed to do that. You're supposed to be pigeon-holed, and you're supposed to fit a pattern.

HKS: And you're supposed to be able to be interviewed. He was reclusive.

CEP: That's right. Now, to give you an insight into maybe the way he thought. You've seen the *National Geographic* article. After the last *National Geographic* article was published, a *National Geographic* person—and I won't call the name—was in our office. He said, "You know, it's a fabulous program, and we appreciate you having us here." We spent a tremendous amount of management time and money providing everything they wanted, whenever they wanted it, so they could do a *National Geographic* kind of story.

And he was expressing his appreciation to Ludwig for having done that. He said, "But I really don't understand how you can justify pouring that kind of money into an isolated region and in a country that you may wake up one day and they have just nationalized it and taken it away from you."

Without hesitating one second, he said, "You must realize that in Brazil that company is mine, and we pursued the ideas that we have. We're doing our absolute best to develop the region because we're convinced that this region, which contributes relatively little today, can make a significant contribution to mankind, and we have the freedom and the ability to pursue those concepts. Yes, it is my money. And yes, it is a risk. And yes, we may wake up one day and it's been nationalized and it's gone. But when that day occurs, I'll know it. And knowing it, I will simply go somewhere else and do something else, because the situation will be defined."

He said, "Contrast that to this country, where you do not have the freedom to pursue these kinds of ideas. You own a company. You're responsible for it. But you do not run it. Government regulations, from whatever agency, of whatever form, decide who you can hire, what you can pay, what you can or cannot do, so that one's economic and mental capacities cannot be expressed. So here you're responsible. You pay taxes, and you don't control it. I choose the other."

HKS: What did Mr. *National Geographic* say to that?

CEP: Nothing.

HKS: Nothing.

CEP: It's hard to respond to.

HKS: Yes, I'm sure it is.

CEP: Because it's so true.

## Leaving Jari for New York

HKS: You left Jari apparently in '75, and the sale took place in '81. Why the gap? Why didn't you stay till the end?

CEP: I was forced to leave.

HKS: Oh.

CEP: By '75, '76, we were at the stage that we knew we were nowhere perfect, but we knew beyond a shadow of a doubt that biologically we could make it work. And so it was time to start the industrial development stage. He came down one time. Of course, we talked about it all the time. But he said, "We have to move." He said, "Who do you know that we can get to plan the industrial stage so that it matches what we're trying to do here?"

He said, "Sure, I can pick up the phone, and I can call Brown & Root. I can call whatever worldwide engineering firm and tell them what I want to do, and they'll take my money and do what I say. But," he said, "I really don't want to do that."

We said we'd work on it. Well, in like a month, he came back and he said, "What's the decision?" I said, "Haven't come up with anybody yet." He said, "You do it." I said, "Look, one of the promises you made when I came to work with this organization was that I would never, under any circumstance, be forced to move to New York City. I'm not interested in changing that promise."

This is one of his abilities of controlling and forcing people. He said, "Fine." At that stage, he didn't say anything. But when I took him to the airport, he said, "Fine. That's your decision, and I respect it, and so whoever I hire and cram down your throat you have to live with. And however they mess up Jari, you have to live with it." I said, "Well, okay." So I moved to Connecticut. Until I left Jari, there was no State-side structure or organization associated with Jari. There was Mr. Ludwig, and yes, everybody in the New York office knew about it, and yes, New York purchasing purchased things on our behalf, and so on. But there was no structure related to Jari.

Once we moved into the industrial development stage, then there had to be structure. When I left, I was still responsible for Jari and responsible for industrial development.

HKS: Was this when you became vice president for the forest products division?

CEP: Yes.

HKS: Okay. Which was Jari, or were there other forest products operations worldwide?

CEP: No. Now, here I am, a forester. Yes, I had had tremendous and probably unequaled opportunities in management, politics, legal, because as manager of Jari, because of his ship's captain mentality, you were legal, political, social, everything, and because of the mining projects and other exploration projects, I had those kinds of experiences, too.

The mining project on the Trombetas had a mining engineer as project manager, but supply, all support, was through Jari. There was just no way to describe or imagine the kind of experience for a young guy. Of course there were a lot of young guys like myself. Whether it was financial or whatever, we got it all.

But still, here we are, entering an industrial development stage, and I'm responsible for it. Yes, I've seen a pulpmill. I've been in pulpmills. I know the paper-making process. I know the chemistry of it. But that's it. So pulpmill, chemical plants, the whole shmear.

I mentioned that one of the means of survival in the organization is knowing when to do, yourself, what he says do, and knowing when to—knowing your own limitations and hiring and getting help in carrying out the objectives that he gives you.

This is probably a spot to step sideways a moment and discuss participation, including contributions, of other organizations. As I stated earlier, this program was not done in a vacuum, as many people assume.

## Joint Ventures and Federal Constraints

CEP: Early on, one of Ludwig's objectives was to have a knowledgeable, experienced, joint venture partner. He did not want to force himself, and us, in a position where we started a world-class manufacturing facility in a vacuum. He previously knew Weyerhaeuser people, and he and George Weyerhaeuser became friends. As a result, we had Weyerhaeuser people living with us for a significant period of time.

Yes, we taught them a lot, but they also taught us. Now, on the average, let's say we had a John Welker, who was fresh out of school with a masters in economics, a brilliant kid. But he's a kid. And Weyerhaeuser in general, when they were sending people to visit us, they were not sending kids. They were sending experienced guys in silviculture, in forest management, in planning. So yes, John was responsible and he had the basic education required to do his job, but not the experience.

But here sat a Weyerhaeuser guy with maybe fifteen years experience, maybe thirty, who was delighted with the opportunity of sitting up till three o'clock in the morning every night discussing technical things with this young guy. So the program with Weyerhaeuser functioned both ways, and because the Weyerhaeuser program came to an end before they got their share, so to speak, the relationship was one-sided, and we got more from Weyerhaeuser for our young guys than Weyerhaeuser got from us.

Now, they got plenty from us, but because of the way the world turns, they were not able, or decided not to, proceed down the tropical path. What they learned from us was not utilized. The moment it was occurring, they were giving and we were giving. It just happened that they didn't get to apply theirs.

HKS: Had they pulled out of Borneo by the time they were talking to you, or was that all happening at the same time?

CEP: Same time.

HKS: Okay.

CEP: The reason some of the people were available to us is that they were leaving Borneo at the time. So Weyerhaeuser was with us during let's say the second half of the forestry development phase. The first half we were by ourselves, but the second half Weyerhaeuser was with us.

About the end of the forestry development stage, the program with Weyerhaeuser ceased, and Crown Zellerbach came in. Remember I stated that Reed Hunt and Ludwig were friends.

HKS: Right.

CEP: Reed Hunt was Crown Zellerbach. Crown had what they called a central engineering group based at Boeing Field in Seattle, and when they decided to build a pulpmill, they designed their own mill. When they decided to build a plywood plant, they designed their own plant. So Crown had their own central engineering to cover everything, all manufacturing facilities, that they built.

When Weyerhaeuser left the scene, Crown forestry came in, and Crown facilities planning came in. This is not to say that we did not have Weyerhaeuser facilities planning people, because we did. But during the Weyerhaeuser stage with us, it was a long-range planning and what-if's for the future. When we were with Crown, it was real because the wood was there and imminent, and there had to be facilities to utilize it.

During this search stage, when Ludwig was trying to find a joint venture partner, we dealt with Weyerhaeuser, with Crown, with Continental Can. With Continental Can it was probably no more than a two- or three-month learning relationship with each other, and they bowed out. We worked with Boise-Cascade in much a similar situation as Continental Can. It was a few-month learning period, and then we stopped.

When we went through all of the forest products companies that we were interested in, and none of that worked, then we started on other companies, including Standard Oil. We worked with Standard Oil probably six to eight months. We worked with ARCO, and we worked with Gulf.

Here's an example of how things that occur in the world alter situations, but I think the first oil company we worked with was Gulf. We were at the stage that all data had been presented, all discussions were complete, and the yes or no joint venture subject was scheduled at the next board of directors meeting, and everything we knew at that stage was that it was a done deal.

That very board of directors meeting—and I don't remember whether you remember this historically or not—but that very board of directors meeting is when the subject of payoffs in international business came up, and the president and chairman, Bob Dorsey, was—it was claimed that he had made payoffs. There was so much turmoil that our subject never came up in the meeting. I don't remember if it was plea bargain or whatever, but that case was essentially the beginning of the large-scale attempts by U.S. government to stop, control, whatever, activities in international business that were deemed to be incorrect, unfair or whatever.

HKS: What was at issue? Lack of regulation caused problems.

CEP: Today, if I were to get back in, which we are in international business, but if I were to get back in international business large-scale, I would not darken the door of any U.S. bank. I would go to an international bank based in the U.S. I'd go to San Francisco, I'd go to Houston, I'd go to Miami.

Because U.S. banks are regulated to the extent that they have no possibility of competing in the international marketplace. Let's say you go back ten years, the largest banks in the world were U.S. banks. I don't know the exact number, but probably the largest U.S. bank today is probably eighteenth on the list. Now, why? It's because of federal regulations that prevent U.S. banks from competing in the world marketplace.

It's so severe that many giant U.S. banks don't even have the know-how anymore to function in international trade. The U.S. Treasury control is so severe and complete in all forms of documentation that, I mean, in American culture and American mentality, it may be correct and fair by our standards. But by the way the world functions, it eliminates U.S. from the competition.

HKS: So an American company like Weyerhaeuser wants to get involved overseas. Is it free to go to an international bank, or does that somehow look like it's a shady deal? To avoid U.S. regulations.

CEP: It would appear to be shady.

HKS: Okay.

CEP: It has a devastating effect on U.S. companies dealing in international trade. [pauses] I'm trying to think of an example. The U.S. Treasury feels that it is responsible to guarantee that Sri Lanka receives its fair share of income from the goods produced in Sri Lanka. So if an unfair price is paid by a U.S. company to a Sri Lankan organization, it's Treasury's responsibility to prosecute. And then it is the U.S. company's responsibility to prove that it's innocent.

As an example, Brazil has a minimum export price on all goods exported from Brazil. That is Brazil's decision. But if that price, established by the Brazilian government, is determined to be unfair in the international marketplace, and you're a U.S. company and you bought it, you're at fault. You get slaughtered by some U.S. agency. So international trade for U.S. businesses is an absolute nightmare.

To continue down this path. We worked with I don't even know the number of companies, searching for a joint venture partner. Almost without exception, everyone wanted to join us. This is my description of why it never worked. From the beginning I told Ludwig that it was futile. We could do it and yes, we could learn something from all of them, which we did. But with his personality, the only way that we could participate with any of these people and reach the objective that he was after in a joint venture was for him to accept a minority position. Almost everyone—Weyerhaeuser, Crown, Gulf, Standard—anyone that we went through the complete procedure with, wanted to do it.

A chief executive officer of a company held by stock-holders was enthralled with it. He was intrigued, delighted, excited with the adventure, but when he came up to the trough, the reality would hit them that they are joining forces with an, in their minds, unpredictable person who, two days after they signed, would decide to plant half of Africa and would drag them by the tail wherever. It simply would not work.

So there was no means to structure an agreement whereby a privately-held, strong-willed company could marry a stockholder-held organization. Because of the magnitude of what we were doing, the importance of what we were doing—we could get people to the trough, but they couldn't drink.

HKS: Why were Gulf and ARCO interested? Because they had a lot of cash to invest? They wanted to diversify? I mean, there was no ultimate energy production involved, was there?

# Early Work with CZ

CEP: They were interested for two reasons. During this stage, they had liquidity. But they also were in the energy business, and the project at Minas Gerais that I had mentioned, which was very large-scale alcohol production, was of interest to them. The *várzea* where we had developed the rice project, ARCO specifically was interested in a large block of *várzea* for sugar cane for fuel production.

During those years, because of liquidity, they were interested in diversification, so the total package of alcohol production for fuel, forestry, aluminum. They were interested in that magnitude of package and that kind of diversification because all of it was natural resources.

Now, with Crown Zellerbach, we had the access to their forestry people and facilities planning, engineering. To drop back in history again, Ludwig and Reed Hunt, back in the I believe late '50s, envisioned and did the initial planning on a floating pulpmill. I mean, when we built the floating pulpmill in Japan and took it to Jari, people thought that was the beginning of the idea. It was not at all, because Ludwig and Hunt conceived of and did initial planning on a floating pulpmill I think in the late '50s, to sit off the coast of Honduras and utilize the pine of that region.

Because of political occurrences, they never did do it, so when it became time for us to plan facilities, those initial ideas and concepts and plans were simply pulled out of the drawer. It was a '50s concept, not a '70s.

HKS: The advantage was that you could manufacture it under First World conditions, with labor and supply and raw material supply, as opposed to building it on site?

CEP: Right. Now, to build it on site, we would have had to have built an additional town for all the construction workers, and you're building under adverse conditions instead of ideal conditions, and a complete package facility—with pulping, power, chemical, everything—built on site, in a completely isolated region, is not easy.

Doing what we did—building towns and railroads and bridges and clearing and planting—is one thing. We did build a kaolin refinery under those circumstances, but the first kaolin refinery is like a ten million dollar adventure in isolated conditions, compared to six or seven hundred million. So yes, we did it small-scale, but to attempt that in an isolated region—and when I say isolated, I'm talking fifteen hundred miles to the nearest industrialized anything. And fifteen hundred miles to two thousand miles to the nearest engineer that can help with anything.

HKS: So Belém is just a place on the map. There's nothing really there.

CEP: It's a city of about, at that time, about a million and a quarter. But it is a city based on a gathering economy. It's not an industrial city.

HKS: I see. So Rio was the next industrial town.

CEP: And Rio is a political play town, and not industrial at all.

HKS: Caracas. Where was your—

CEP: São Paulo.

HKS: São Paulo.

CEP: Which is further south than Rio. So if you want a water pump, it comes from São Paulo. If you want plastic pipe for plumbing, it comes from São Paulo. The combination of factors, and the fact that financing was available through the Japanese, it was a shipyard where Ludwig had built all his ships, it was.... And the difficulties of doing it at Jari. It was a monumental undertaking, but it worked.

All of the initial engineering was done by Crown. Essentially we built a mill that was a carbon copy of an existing Crown facility. It's no longer Crown today, but it was an existing Crown facility. So we weren't trying new concepts. It was a mill that had been running for years and yes, where they knew there were problems, there were improvements. So the mill was absolutely modern, up-to-date. But it was not a wild guess deal out in the jungle.

To carry this another stage, once we realized that we were not going to have a Weyerhaeuser or a Crown with us, then another employee and I—his name is Gordon Douglas, who is from Vancouver—went to Finland and developed a relationship with a Finnish company. We signed an agreement whereby that Finnish company would come live with us during construction and startup, and train all of our Brazilian employees.

So we sent shiploads of gmelina to Finland. We hired all of our mill operating people and management, and sent them to Finland. So they had the mill startup, with gmelina, in Finnish mills, with the supervisor that was going to accompany them back to Brazil. So at startup time, we had a complete complement of some of the best, if not the best, pulp people in the world running our plant.

And they stayed there. It was not a given date that they would leave. It was man-by-man, section-by-section, so that if the Finnish guy responsible for the digester said, "Look, this Brazilian guy, he's fabulous and he's learned everything I know, I'm ready to go home."

The Finnish contingent left one-by-one, rather than en masse on a given day. I don't know all the mills anywhere, but I know that our mill's startup was probably—even though in its isolated condition—one of the easiest and the quickest startups in attaining of capacity of any mill around. The mill has been operating above capacity ever since.

Then, on marketing, we went to a London worldwide marketing firm, called Price & Pierce, and we contracted with them to do the marketing. We blood-and-gut dreamer pioneers of Jari knew better than to try and start up a pulpmill on our own hook and knew better than to try and do world marketing. Independent, new company, no one ever heard of you before, no one ever heard of the fiber before, and pulp is a commodity.

One of the reasons that it has functioned efficiently since day one is that the market existed because of Price & Pierce, and the quality and quantity has been good because of the Finns.

## Swiss Institute

CEP: One of the things that probably hurt Ludwig more than all occurrences in the years that I was involved was the foundation situation. The subject of organization and structure was always on the table and could never get resolved, and because it couldn't get resolved, the only alternative was a foundation and elimination of all of his operating assets, and to turn those assets into paper assets that can be managed in an office and does not require Ludwig.

So those kinds of discussions were around and ongoing from the day I got involved. He and everybody realized the problems and importance of it. His hope and dream was to establish a U.S. foundation. At the time, it would have been probably the largest U.S. foundation in existence, and a free gift. Guess what? Our government would not accept, would not allow the American people to accept a free gift.

The bulk of his assets were worldwide. So a coal mine that he owns in Queensland is an offshore asset. He may take a dollar from that, and put it into an office building in Honolulu, whatever he wants to do with it. And so, a dollar he makes on that coal mine—I shouldn't have said Honolulu, because that's U.S. But a dollar he makes on the coal mine, put into an office building in Sydney, is not taxable in the U.S. But when he sells all of these assets and brings those dollars to the U.S. to establish a foundation, it's all taxable.

HKS: Aha. I see the problem.

CEP: So you take billions of dollars in assets and bring it to the States. There's nothing left for a foundation. That was fought and fought and fought. So one of the hardest things he had to do was admit that he could not give to his own country a gift. It had to be offshore. And that is a terrible comment on our system. But there was simply no way around it.

HKS: The articles I've read said the Brazilians had a real problem at Jari because it was going to be owned by an outfit in Switzerland. And this bothered the Brazilians.

CEP: That's a figment of the media's imagination.

HKS: So the problem was really Ludwig's personal problem in terms of what he wanted to do, he wasn't able to do. But in terms of Jari, it didn't have an impact.

CEP: No. Bottom-line decision point on the subject of Jari is social responsibility and the government's inability to pick up and carry the load that they had agreed to carry.

HKS: So he picked Switzerland, because obviously the tax situation is more advantageous.

CEP: Correct.

HKS: And it was dedicated totally to cancer research, this foundation? This is what I've read.

CEP: Yes. Funds from the Swiss foundation can come into the U.S. to U.S. organizations doing cancer research, and not be taxed.

HKS: Oh, I see. So they're not doing research in Switzerland. That's just where the money is. That was never spelled out.

CEP: So it's Swiss simply because the asset base can be in Switzerland without being taxed, and therefore the revenues can be distributed worldwide to other foundations without being taxed.

HKS: Okay.

CEP: But if you put a U.S. type tax load—state, federal, everything else—on a gift coming in, then the money left that can be distributed for medical research is diminished dramatically.

## Agricultural Crops

HKS: The speculation that the world had of Jari was that it was an experiment to see if development would work in the Amazon. Somebody wrote that seventy percent of the Amazon could be developed. I don't know what the other thirty percent is. Too mountainous or something. In a sense that was Ludwig's plan, too.

CEP: Well, they're probably assuming that the part that can't be developed is savanna or *várzea*. But contrary to what most people think, probably the best and the most suited and easiest land to develop for agricultural purposes is *várzea*. Generally—and this is one of the concepts that is used that is not very valid—agriculture is not one of the ways to develop the Amazon.

We spent I don't know how much money—low millions—in trying to develop agricultural crops. Cotton is a tree, and it's native in the Amazon. We have taken this tree and genetically reduced it to a little bush that can be managed on an agricultural basis.

HKS: I didn't know that. How about that!

CEP: There are cotton trees at Jari, and we tried to grow cotton. And yes, you can grow cotton. But when you have sufficient moisture year-round and no dormant season, by the time that it decides to flower and provide some cotton, you have a tree.

The problem with soybeans. We grew soybeans. You can produce some of the most fabulous soybean plants the world has ever seen. But plants are triggered to flower and reproduce, produce seed, based upon environmental circumstances. Of course, a lot of activity in plants is controlled by photoperiod. So if the mechanism of photoperiod for a given plant doesn't occur in the tropics, it doesn't flower. If you're growing forage for cattle, you can produce tremendous amount of vegetation in soybeans, but you don't get much of a seed crop.

The standard thing that you see on TV is land-clearing, burning, and then big agricultural tractors preparing the land, and somebody growing soybeans. Rest assured that that's not in the Amazon basin.

HKS: It's farther south.

CEP: It's south. A horrible comment, and one of the things that as a newcomer to the Amazon was extremely difficult for me to believe, but in the market in Belém, you can get fish, herbs, and everything you can imagine, and fresh fruits and vegetables, and whatever. Guess what? They're grown by Japanese in southern Brazil. Every airplane daily flight had fresh fruits and vegetables that we bought in Belém, but they came from southern Brazil.

There's cashew, there's banana, there's pineapple, there's papaya. There are all kinds of fabulous and delicious and nutritious foods in the Amazon that native people eat and that I prefer above anything else. But in terms of foods that most so-called, quote, "civilized" people eat, they're not grown there. Growing almost anything there is extremely difficult.

We grew fryers so that we had chicken. We produced bananas, papayas, all kinds of things for people to eat. Of all activities that were the most miserable, it was trying to produce vegetables for the table. We had tropical agronomists hired specifically for that purpose, and never really succeeded. Yes, we could occasionally get a tomato. And yes, we could occasionally get a leaf of lettuce. But—

HKS: Is that worldwide on the Equator? The problem because of photoperiod?

CEP: Yes. That's not the only reason. Things that we normally consider as necessary food items are not tropical plants. They did not originate there. They were not selected and bred to function in that kind of environment. So for people to have fear that the Amazon is going to be cleared for large-scale agriculture is not valid. I mean, even if there were varieties suitable, the soils are not suitable.

Rainfall is a horrendous problem. Name me anyplace on the face of the earth, including valleys in California, where tremendous volumes of foods are grown. They irrigate, but they would have a wipe-out if they had a hundred inches of rain. I mean, a hundred inches is an inch every third day.

So you have horrendous pathological problems with almost any garden plant. The only way to get a tomato is that there are distant relatives of tomatoes native to the Amazon. The fruit on them is not edible. The plant is thorny. It's a very unlikable plant. Because they're related, you can graft a tomato onto the root stock of that plant. That plant you can graft on is taking the brunt of the attack of the natural world of pathogens and insects and whatever, because it's in the ground, and the part that's producing the tomato you're after is the aerial part of the plant. So you have a much better chance of getting a tomato.

But the problems with agricultural production as we know it today are so daunting that there are too many other places in the world that it's too easy to grow something to eat on to mess with the Amazon. And yet people think it's the opposite.

HKS: What is the potential for developing the Amazon for the Brazilians, other than what's going on: land clearance for small-scale agriculture?

CEP: The natural forest itself produces a tremendous volume of items that are usable.

HKS: Non-wood items you're talking about.

CEP: Well, wood and non-wood.

HKS: Okay.

## Ludwig the Man

HKS: Let's return to Ludwig the man.

CEP: Ludwig's wife is named Ginger. She was a red-headed Irish girl just slightly younger than he. She controlled his life much more than anyone dealing with him on a daily basis would imagine, because you would think no one would control his life. But any Christmas present, birthday present, anniversary present that she ever gave him, he'd grumble about it, wasted money. He didn't need another shirt. He'd take it to the office and have someone return it and get his money back. If you have thirty shirts in your closet, you only wear one at a time, what's the logic of another one? Obviously, if he needed a shirt, he'd buy one. But to spend money on a shirt when you had thirty simply did not fit his pattern of thinking.

She was participating in the building of a children's home somewhere in the region. As I understood it, not far from their apartment. She and he were contributing the landscaping. After dinner one evening on his birthday she took him by the arm to show him the progress they were making in landscaping for this children's home. They go out into the area being landscaped, and she said, "Here is your birthday present." It was two truckloads of chicken manure to be used in establishing the plants. She said, "If you don't like this present, you take it back yourself."

HKS/CEP: [mutual laughter]

HKS: Did he have a sense of humor?

CEP: Yes.

HKS: He got a kick out of that, I'm sure.

CEP: That one was so delightful to him that he relayed that to all his friends.

HKS: That's interesting. Was business really all he had in his life?

CEP: Socially he mixed with the Reagans. Clark Gable was one of his closer social friends. Socially he enjoyed the fact that he had power and could associate with those people. What do you call it, the Bohemian Club?

HKS: Yes.

CEP: Where these powerful guys occasionally get together. Well, he was part of that group. But socially, he was definitely more business than anything else, and his social activities were, in one form or another, really related to business.

HKS: But if he liked Clark Gable he didn't buy a movie studio or something.

CEP: No, no, no. Let's say he was trying to get a large U.S. corporation interested in a mining or oil or whatever kind of venture on the other side of the world. He would call the chairman, and the guy would already know about the program and be following it. Ludwig now has it to the stage that it's time to go big-time in development.

He calls the guy and says, "We're ready, and we need to go look at it and really determine if you're interested in participating or not." The guy says, "Fine. I'm ready to go." Ludwig says, "I'll meet you at the ticket counter in New York, and we'll go together, and we'll discuss it as we go. From there I need to go"—let's say they were going to Africa—"From there I need to go to Brazil, so we'll part company in Nairobi."

The guy knows that Ludwig flies coach and knows that he doesn't need any of the so-called luxuries that this guy has to have, so within an hour the guy calls back and says, "You know, I've checked and our corporate jet is not doing anything next week. So we might as well go in it." Ludwig says, "Well, whatever you want to do. This trip is for you, to show you what we're doing. Whatever you want to do."

Well, that saves Ludwig a round-trip ticket to the other side of the world.

HKS: [chuckling]

CEP: Time and time and time and time again [chuckling], the cost of doing things he would accomplish in getting it done without bearing the cost, and without the other person ever knowing that he paid for it.

HKS: But if he had to buy his own ticket, he was reluctant to do that.

CEP: Oh, no.

HKS: Did he like to travel? Or was that just part of his job?

CEP: It was part of the job. He traveled a lot, traveled easy. He had the ability to prevent things beyond his control from bothering him. If it was something he was supposed to be able to control, he could go ballistic on you. But he could have a meeting the next morning with the most important guy in the world, and if a plane was broken down, it was broken down.

Things beyond his control just simply he didn't allow them to enter the equation, because if he got upset because the plane was delayed and he wasn't going to be able to make the flight or make the meeting, and he got upset and frustrated about it, it would steal from him his thinking time.

HKS: I can understand that. Something more of us ought to pay attention to.

## Projection of Supply

HKS: That says a lot about Ludwig and his values, making use of the moment. Let's go back to projection of supply.

CEP: The subject is the raw material for the pulpmill. Is that not it?

HKS: Yes, that's it.

CEP: At day one, your projections are mere speculation. Each day that goes by, as you collect data, the projections become more valid.

Not knowing exactly what would happen, we started making contingency plans. One of the things we did, which I've really never heard anyone else comment about, but we started and we (gmelina is a short-fibered, white-colored wood) identified about four hundred and sixty native species. Out of those four hundred and sixty, we identified and tagged, so to speak, all species that were white in color and/or were at least light-colored and could be easily bleached, because gmelina was to be a market-bleached pulp.

Any species that conceivably could be mixed with gmelina we collected, and I took samples to either Herty Foundation in Savannah, Institute of Paper Chemistry in Appleton, or N.C. State, and did pulping studies. Because, number one, if those species can mix with gmelina and we're clearing land to plant pine, gmelina, and eucalyptus, what is the logic of burning those species when they could come to the mill.

At that stage we did not know what the logical rotation period should be. If you could feed the mill with a twenty percent mix of white woods from the rainforest that you were burning, that means you conceivably could increase your rotation period a year, if you needed to. So if the mill is there, functioning, and you missed the calculation of what the biological and economic rotation should be, you could extend the rotation and still run the mill full capacity.

We identified all the species, upland and lowland, that would meet that criteria. As it turned out, the volume of those species was significant. Planting of pine in the Territory of Amapá started in earnest probably in I guess '71. I need to back up a little bit. There was an FAO project in the Amazon and I've forgotten the years but probably late '50s, early '60s. It was mainly tropical species, but they did plant *caribaea* scattered around in different places, and one of the places was in the savannas, where we started planting in '71.

They weren't maintained. They were abandoned. But they were surviving, and they were there. So at least we had an idea of what a twenty-year old tree in a savanna would look like, same species. Had no idea of the seed origin or anything else, but at least there was an indication there.

At the time the Jari pulpmill started, pine was available from the plantings in Amapá. Remember that the owner of the plantation in Amapá is now the owner of Jari.

HKS: Okay.

CEP: To an outsider who doesn't know, he sees wood being cut and put on barge and hauled to Jari, and so Jari doesn't have enough wood. He sees jungle species on a barge going to Jari, so Jari doesn't have the capability of growing their own trees. One of the things that did happen that contributed to the use of native wood and Amapá pine is that the last year or two that we were at Jari, when we knew that we were leaving, but before there was change in ownership, the money available for plantation maintenance decreased dramatically.

That was a major effect on growth. Remember I said that growing plants there is like tomatoes in your backyard. If you do not do what is supposed to be done on schedule, tomorrow you know, based on growth, that you didn't do it. It's not consequences that you pay after you retire. It's consequences that you can see. You don't have to have a diameter tape. You don't have to have an altimeter. You can drive down the road and see the difference between what you did on this side of the road and what you didn't do on the other side. That was the beginning influence.

The second thing was that new management were mining people. The owner was a miner. The new manager was a miner, and with a mining concept, you utilize what you have. What you have is a given. But the difference between a deposit of manganese and a fast-growing, sensitive organism is very great. So, number one, the management of this organism disappeared in the change. The money for caring for this organism disappeared.

There was a critical period of three to five years that they had to learn the reality of the kind of tiger that they had by the tail. There was a dramatic influence on growth rates. Without material from Amapá and without the planning we had done on native species, they would not have been able to operate the mill full tilt.

Growth is not below projection. In reality silviculture was dramatically less than projected activities thus contributing to less volume growth. Our pine growth, based on today's growth, is about fifty to sixty percent greater than we projected, and we were projecting unbelievable numbers. In fact, we had trouble with them because they were so high we couldn't believe them, but there were the numbers. The actual is greater than what we projected.

As I mentioned earlier, Jari management has done a fabulous and beautiful, fortunate turnaround, and they now fully realize what they have and are doing. I have not been there recently, but people that have been there say that they are carrying out silvicultural practices that are giving excellent results. Of course, that's part of the reason that growth rates are higher than what we originally had projected.

HKS: I guess people just wanted it to fail.

CEP: That's right.

HKS: That you can't conquer the Amazon.

CEP: It can't be done, and we don't believe it can be done, and so the fact that you did is immaterial. It still failed.

HKS: I don't know anything about major developments in Africa.

CEP: South Africa is an exception. South Africa has large-scale, beautiful plantations, mainly pine. Well-managed, productive. But in tropical Africa, I won't say everything because I don't know, but everything I know about failed. Not necessarily for any biological reason but because of lack of government continuity, political continuity. The British might go in and start something and do it right, and it's progressing and it's beautiful, and they leave, and the government changes, and the next government doesn't allocate one cent to the maintenance of what was established, and in a short time it doesn't exist. It exists in a record book somewhere, and that's it.

For example, we collected seed in many countries in Africa. Nigeria, Malawi, what was Rhodesia, everywhere that they planted. In many instances the soils were fine. There wasn't a thing wrong, except that there wasn't a single area that was successful because management was zero. You can't grow tomatoes with zero management.

HKS: So Ludwig was correct in not putting Jari in Nigeria. The instability would have prevented his success.

CEP: Right.

HKS: That's a good place to end. Thank you very much.

Laying railroad line. Welker Photo.

# ROBERT J. GILVARY

## Introduction

Robert John Gilvary was born on May 17, 1938. His resume dated February 29, 1984 states that he is an American, but a "resident of Brazil since 1962, with an exception of two years." He has since returned to the United States and owns and operates a hardwood tree farm near Blacksburg, Virginia. In fact, he enjoys the tree farm business so much that it was difficult to schedule the interview that follows. He joked by phone that he would be available on any day that it rained. As it turned out, it was clear the day of the interview, but it was also the day before hunting season opened, so we didn't dawdle—we had to finish that day.

The interview took place in his home office at the edge of Blacksburg. The office looked much the way one would imagine an engineer's or tree farmer's work space would appear, with files, charts, and reports filed, stacked, or spread throughout. His Brazilian-born wife Darlinda made certain that we were well supplied with coffee, and sandwiches, too.

It was Bob's Brazilian period that we focussed on, especially the time he was employed by Universe Tankerships and its forestry project along the Jari River, a major tributary of the Amazon. Daniel Ludwig, the legendary owner of Universe Tankerships, personally interviewed Bob for an engineering job.

Bob's assignment would have daunted most; he would eventually build four towns, four hundred kilometers of truck road and forty-four kilometers of railroad, a 120-acre nursery, foundations for two forty-thousand ton platforms holding a pulpmill and a power plant, a kaolin processing plant, a sawmill, and a maintenance complex. Oh yes, a sixteen kilometer power line and an airport were also on his list of things to do. Part of his task was to oversee the fourth largest fleet of heavy equipment in all of Brazil, consisting of 240 bulldozers, backhoes, scrapers, haulers, and on down the list. In fourteen years, this equipment would move twenty-five million cubic meters of earth.

# Hired by National Bulk Carriers

Robert J. Gilvary (RJG): I was working with a contractor in Pennsylvania, building a stretch of interstate highway. After I'd been there close to a year, a friend of mine in Belém wrote me and told me about this big project that was starting up in the Amazon, an American owned company. She gave me the name of the company, National Bulk Carriers.

I found out where their headquarters was, and at that time it was on Lexington Avenue in New York City. I drove to New York. I walked into the office and said, I hear you have a big project in Brazil. They said, Yes, we do. I said, I'm a civil engineer and I'm interested in working for you. At that time I hadn't the slightest idea who Daniel Ludwig was. I'd never heard of him. I didn't know what National Bulk Carriers' business was. Anyway, the guy I was talking to was the guy that Ludwig had responsible for the project at that time, a guy named Frank Penn.

I guess Frank liked me, and we got talking. He said, yes, we do have an opening. It's a forestry project where we need to establish a two hundred fifty thousand acre plantation of this fast growing species called gmelina. He explained to me something about what the project was about. Then he said, Everyone that goes to the project Mr. Ludwig interviews. Mr. Ludwig is the owner of the company.

Harold K. Steen (HKS): Is that right? He actually interviewed you?

RJG: Yes. He said, Let me send you in to talk to Mr. Ludwig for a while. Again, I had no idea who he was or.... [both laugh] So I went in.

HKS: Describe your first impression of the man. I've read about him in newspaper accounts, what he looked like and so forth.

RJG: He was seventy years old at the time, and he's rather ordinary looking. The first thing he said to me was. Well, Gilvary, I see you were in the Peace Corps, eh? I said, Yes, sir. He said, Didn't spoil you, did it? I said, No, sir, I thought it qualified me even more for projects like the one you're working on in Brazil. We talked for a few more minutes. I don't recall what else was said. And that was it. The interview with him lasted perhaps five minutes. Then I went back out and talked to Mr. Penn, and a few days later I was offered the job, at a lower salary than I had asked, but I jumped at it anyway. I think I started at eight hundred dollars a month, or something like that.

In the meantime it was Frank Penn's secretary, Miss Antonucci, explained to me that Mr. Ludwig [laughing] was the richest man in the world, and what his business was. He owned the largest fleet of supertankers in the world, he was essentially the man that had invented the supertanker, and that he had about, at that time, probably sixty different companies around the world. I said, Fine. Two weeks later I was off to the project in the Amazon.

## Early Situation at Jari

HKS: That's how you got started. You arrived. Clayton Posey went into some length about conditions on the ground when he got there. You were ahead of Clayton?

RJG: Much earlier. The project had only been going a few months—

HKS: Clayton describes it as a project run by a bunch of engineers. So you're one of those guys.

RJG: I was the first engineer, but there was a forester there who had established the first nursery, and who had seedlings growing when I got there. His name was Tom Bunger. He was the original forester, the only one. His responsibility was to establish the nursery and plant and grow the seedlings to plant the twenty thousand acres that was supposed to be cleared the first year. As it turned out, much less than that was cleared.

HKS: It must have been relatively primitive then. I mean, the infrastructure hadn't been developed. That was part of your job, to create a town.

RJG: My job was not defined. At that time, no one from Mr. Ludwig on down really had a vision, I think, of what the project was going to develop into. The objective was to clear twenty thousand acres and plant it to gmelina this first year. No one thought about building a town, or roads, or railroads, or anything else. I mean, this was the objective. When I got there, it was just a small construction camp, really. You know, some primitive housing, no family, everyone was there on bachelor status. A small landing strip had just been completed for single engine aircraft, and this was their means of communication with the outside world. When Mr. Ludwig bought the land, he actually bought out the Brazil nut business that the former owners had on the land. This included several riverboats, one of which was a substantial vessel, two hundred feet long and diesel powered. It could haul about three hundred people at a time from Belém to the project. And this is how most [emphasis] people got there, workers other than supervisory personnel. We flew out and back in small rented airplanes. The Cessna 206, a six seater single engine aircraft, was the aircraft we used in the first few years.

# Daniel Ludwig

HKS: I thought there was a master plan. Obviously not. You called him Mr. Ludwig. Did you know anyone who called him Daniel or D. K. or anything?

RJG: [slowly and emphatically] Heavens, no. [both laugh]

HKS: Would he have been offended? Or just his bearing was such that you wouldn't even think about it? Or the fact that he was so rich, I suppose.

RJG: He's probably the only person that I refer to as Mister. He commanded that respect, let's put it that way. His personal friends, those of other wealthy men who were his friends, would call him Daniel or D. K. However, I was never in their presence. My contact with him, obviously, I was always an employee. And even when he was there with other friends of his, I never heard anyone call him D. K. or Dan. [both laugh]

HKS: How did you know what to do? Who was your supervisor?

RJG: At that time the project manager lived in Belém, and there was an on site superintendent. The project manager at the time was—and he was not the first, even he [emphasis] was not the first one. I don't know if Clayton got into how many project managers...

HKS: Thirty of them or something.

RJG: Right. A lot of them went through. But at the time this man here [photo] was the head of the project, living in Belém. And this guy here was the on site superintendent.

HKS: So he came up and back the same day wearing a suit and a tie, right?

RJG: There he is there in a coat and tie, which explains why he didn't last very long. [both laugh] Like so many of them. As it turned out, when I was hired I flew down to the project with the head guy. We happened to be on the Pan Am flight out of New York on that same day, and he barely acknowledged my existence. He was Danish and he had a wealth of experience. He was more of a diplomat than anything, and with Ludwig that sort of personality led to disaster fast. With Ludwig you had to be concise and to the point. He had no patience for diplomacy. It eventually was his undoing, which I'll get to ultimately. He wanted people that got things done and that didn't give him any bull. And those are the only ones that survived on the project.

HKS: I like you coming back to Ludwig because he's sort of a fascinating character.

RJG: Yes.

HKS: In terms of micromanagement, did he only talk to the director? Or when he was there, he walked up to you and asked you what you were doing? Did he know your name, and he knew you were an engineer and asked you directly to report? Or did you go through some hierarchy?

RJG: He essentially paid no attention to most of the general managers. When he came to visit the project, if it were a forestry subject he would talk to Johan Zweede, and if it were a construction subject he'd talk to me. I've got a bunch of memos here from him which will indicate to you the detail this man got into. [laughing] I kept these because I think they're priceless. There are things there you will not believe. Here's a billion dollar project and the owner, who is the richest man in the world, is communicating with someone like me on subject matter like this.

HKS: Give me an example of this, trivia obviously.

RJG: Let me pull one out of the hat here. To R. J. Gilvary from D. K. Ludwig, December 30, 1976: Enclosed find a letter from Mr. Buford of Hannah Mining Company regarding tests on stabilizing material that you propose to use in our industrial section (he means for the industrial site or the pulpmill site). I suggest we follow through by an application of clay on top of the sand and also by mixing the sand with the clay. Indications are that this should materially improve the feasibility of using this material. You will note that by applying the material to stiff clay, the material is moved up from one ton per square foot of soft clay to three tons per square foot of stiff clay (what he means is that the load bearing capacity goes from one ton per square foot to three tons per square foot). By putting a balance of stiff clay and rolling it to a blanket of six inches, it should not be too exorbitant in cost. D. K. L.

HKS: How does he know that? A report?

RJG: This friend of his from the Hannah Mining Company told him this.

HKS: O.K.

RJG: They just go on like that, you know. The guy was something. [amused, but admiring]

HKS: Did he—I don't know how often he came down—but did he really follow through? I sent you that memo back so and so?

RJG: Oh, yes. You'll see in here that several times he chews me out for not responding [both laugh] to a previous memo.

HKS: Incredible. Incredible.

RJG: Yes.

HKS: When you think of all the other companies around the world he owns—

RJG: Although at the height of the construction of the pulpmill, '76 through '78, he probably spent, I'm guessing, eighty percent of his time on the Jari project.

HKS: O.K. But still, if there wasn't a plan, how did you know what kind of roads to build? What kind of truck traffic there would be, years in the future?

RJG: That's what I started doing, now we had an engineer there. The original objective was to establish a two hundred fifty thousand acre tree farm of gmelina—nothing else. This fast growing exotic species is native to Asia and we had seeds imported from Africa. The objective was twenty-five thousand acres a year, or ten thousand hectares, over ten years to establish the plantation of two hundred fifty thousand acres and then industrialize. In the beginning it was not defined what was going to be done with the fiber from this tree. There were talks of a pulpmill right from the beginning, a hardboard mill, sawmill of course, possible plywood mill.... I'm guessing now, but I don't think a decision was made to actually build a pulpmill until '74 or '75.

HKS: Was this perhaps prudent on his part—he wanted to make sure the plantations had adequate yield, or...? What would have been the market if he had never built a mill?

RJG: I think his original concept was that there would be a severe shortage of wood fiber in the late 1980s, and that whoever had the wood fiber would make a bundle. That's what prompted him. I think pulp and paper were always there, but not necessarily as the prime product or the only product. First of all, he thought the tree would grow much faster than it did, so he envisioned saw timber and actually peeler logs for plywood.

HKS: Clayton said consultants just rained on him. This is the way he described it, Ludwig was always sending consultants down.

RJG: I can understand in the forestry aspect of it, Ludwig was always sending someone to second guess you. For every one I had to deal with second guessing me, the forester, either Clayton or Zweede, had ten to contend with. I had my problems with consultants he sent there, because people started second guessing me.

HKS: That was just Ludwig's choice? Why did he decide that forestry needed that help, and not the engineering part?

RJG: It took many years for me to gain Mr. Ludwig's confidence. However, once you had his confidence he gave you relatively free rein.

HKS: I see.

RJG: Like I explained before, I kind of grew with the project. We started out as a very small construction project. I grew with it. I was only twenty-nine when I hired on with Ludwig, and he was seventy. My age was against me, and also the fact that I had a college degree was against me. He had contempt, really, for people with college degrees, although he recognized that certain things required it. High tech engineering, he knew that you couldn't build a rocket to the moon or jet aircraft without engineers. But he still had contempt for people with degrees.

But I had my share of consultants, and, for example, just take simple dirt moving, right? I would order, or put in the budget, the heavy construction equipment I needed to do what had to be done, the grading for the railroad, the building of the main road system, etc. And I'd get a consultant or someone that Ludwig would hire and come down and say, What do you want that for? In California we don't use Cat 631s, we use 651s. You get a lot more production. You said, Well, maybe that's how you do in California, but here in the middle of the Amazon it won't work. And I was fending off stuff like that all the time, too. Don't get me wrong. But I, you know, I generally managed to prevail.

HKS: How long does a piece of heavy equipment last when it's used a lot?

RJG: We went through two complete sets of heavy construction equipment in the thirteen years I was there.

HKS: So you originally started with D-8s and D-9s, and you downsized?

RJG: The D-8s and 9s continued, except that we no longer used them in clearing. They all eventually were used in construction. We started out with a relatively small type of earthmover, a scraper, for moving dirt, and we upgraded one size on the next go around with them. About the time we were getting into real heavy earthmoving, where Caterpillar could not deliver on time, they were swamped with orders and we had to switch to Terrex, which was somewhat inferior to Caterpillar, but nonetheless they got the job done. More than anything I was second guessed on the type of equipment that I wanted, because Ludwig was fascinated by equipment, and especially exotic equipment.

HKS: Did salesmen come down? Seems like you were a pretty good market.

RJG: No. No. Because of the remoteness of the project, and the only feasible way to get there was by company plane. A salesman could have hired an air taxi in Belém and flown out there, but he gets there and he's on his own. He's got no car, no nothing.

HKS: I realize there's all sorts of technical specs that are published about equipment. But how did you gain your experience, so that, when you wanted to make a shift in the technology, that you knew the new equipment was going to do something better than the old equipment?

RJG: I didn't really require any exotic equipment. It was standard earthmoving equipment. Bulldozers, and scrapers, and hydraulic excavators, and compactors, and off-highway dump trucks. For the rock quarry, I went to the best equipment—Gardner-Denver compressors and drills, and the rock crusher. Ludwig and I went through a long process on that before we finally got a rock crusher that I needed to do the job.

For example, Ludwig wrote to me. He's telling me that I'm becoming overworked and they're asking too much of me, so he's decided to divide the engineering construction division into two parts, where I would be responsible for heavy construction and someone else would be responsible for residential and lighter construction. It never came about. [laughs] Nothing ever came of this.

HKS: He didn't do that as a way of demoting you.

RJG: No.

HKS: He would tell you point blank, right? He wasn't shy.

RJG: Oh, would he ever! Ludwig could be ruthless. There were several times where, with two general managers in particular, that he interviewed and sent to the project to be general manager and I was reporting to them, forestry was reporting to them, and maintenance. And the first time he comes to visit the project, he treated these guys like dirt. I mean, right in front of us, he would ignore them. We were at a meeting, right? These poor guys would try and talk to him, and he wouldn't even talk to them. He'd talk to me if it was about construction, or Johan if it was about forestry. He just got turned off by these guys, something they had said or some wishy-washiness that he noted. He could be ruthless in that respect.

On the other hand, Ludwig could be very understanding. He was very understanding when it came to something like illness, especially serious illness. I don't know why he was so concerned about cancer, but you probably know that he donated most of his wealth, he formed the Ludwig Cancer Foundation, which is today administered by a group he designated. But he was very understanding about something like this, and he would intervene personally. My mother got cancer in '77, and the local doctors in the local hospital couldn't do anything for her.

HKS: She was in the States.

RJG: My mother was in the States, right.

HKS: O.K.

RJG: So it's the one time I talked to Ludwig that was not about business. I called Mr. Ludwig from Monte Dourado, and I said, Mr. Ludwig, my mother has cancer, and the local doctors can't do anything. I wonder if you could get her into the Mayo Clinic. I always choke up when I tell this story. But at that time, I think there was like a six month waiting list to get into the Mayo Clinic, and she was in there within a week.

HKS: Interesting.

RJG: I never heard the results of her stay. I didn't hear until the next time that Mr. Ludwig visited the project. He just said, Bob, you know, you only get one chance with cancer. And that's about all he said. I knew that at that point it was just too late.

HKS: He called you Bob? Did he generally use first names with people?

RJG: I was Gilvary until I gained his confidence.

HKS: I see. He's certainly a fascinating guy, and the U.S. press had a lot of fun stereotyping him as being a reclusive.

RJG: He was not reclusive at all. Comparing him to Howard Hughes, for example, is ridiculous. Howard Hughes hid—you couldn't see him personally. Ludwig was not reclusive. If it concerned business, he'd go anywhere, talk to anyone. He just didn't want any extraneous activity, like reporters or salesmen, or that, getting near him. Like I said, he visited the project every three months during the peak construction of the pulpmill. He flew—initially he flew tourist class, and that's another one of his aspects. He would not waste money. Then in later years, when his back was bothering him real bad, he switched to first class.

HKS: O.K.

RJG: He'd fly Varig from New York to Belém, or Miami to Belém.

HKS: When you took him out on the construction site, in undeveloped areas, he'd sit there leaning against a pickup truck and eat a sandwich? He didn't need any special food or catered...?

RJG: No, none whatsoever.

HKS: He got a little dirty and that was all right with him. He didn't worry about that.

RJG: Yes.

HKS: He didn't come out with a suit and tie on?

RJG: [laughs] Never. Never. He dressed just as casually as could be. If he wanted to go to the rock crusher, he'd say, O.K., Gilvary, let's go out to the rock crusher. He'd hop in my pickup, and we'd bounce out to the rock crusher.

HKS: You said earlier you felt he was spending seventy-five percent of his time on the Jari project.

RJG: Yes.

HKS: Because he really got fascinated with it, apparently.

RJG: Well, fascinated, and it became far and away his major investment. I mean, even though he had something like, at that time, about ninety companies around the world, this is by far the biggest project, the biggest undertaking of his life.

HKS: I don't want to get ahead of the story, but would you have done a lot of things differently had you known what ultimately would have been the size of the infrastructure? Or did it actually work out pretty well?

RJG: One mistake that was made—which was irreversible even when I got there—was the location of the town of Monte Dourado. It should have been next to the pulpmill to avoid the sixteen kilometer commute that everyone who worked at the pulpmill and the industrial site there had to make. Other than that I don't think there's a major disadvantage in the lack of planning. I mean, there were adequate town sites next to the pulpmill.

HKS: O.K.

RJG: Monte Dourado was kind of just picked randomly as a construction site. People waded ashore there, at what came to be Monte Dourado. The first nursery was established there. Even by the time I got there, with the construction camp plus the nursery, you had considerable, significant infrastructure already there.

## Early Assignments

HKS: What was your main job? Roads?

RJG: We conceived of what the project needed in the way of construction, then we designed what we conceived of, and then we built it. We didn't use outside contractors. Everything was done in-house. It was a unique opportunity for an engineer, really.

HKS: What sort of approval process did you have? At each step of the way? Or random? How did it work?

RJG: We prepared budgets each year. The budgets were seldom followed. It was obvious that we're establishing this two hundred fifty thousand acre tree farm. We needed to have all-weather roads just to get out to plant the trees. Obviously, we're going to need these all-weather roads to bring the wood to whatever processing plant we build. This is one thing I [emphasis] pushed for right from the beginning, let's lay the roads out in at least where we think they should be to ultimately get the wood to the mill. Maybe we don't have to make them as wide as we will need to eventually, but let's at least put them in the right place, so we can at least widen them later. That kind of helped to define the road system.

HKS: Did you have a quarry and a crusher, the whole works?

RJG: No. At that time we didn't even know there was rock there. This being the tropics, we were blessed with some naturally good road surfacing materials. We had river run gravel in some places, and we had laterite, both iron based and aluminum based laterite. The iron based laterite was an excellent road surfacing material. So we never actually had to use crushed rock on the roads.

HKS: There were no so-called environmental constraints on getting gravel out of the river?

RJG: We didn't get it out of the river. These are deposits far from the river. And at that time, in the late 60s, early 70s, there were no environmental constraints anyway.

HKS: O.K.

RJG: I don't want to get ahead of the story. We eventually found a very good deposit of rock, a very hard diabase rock, which turned out to be an excellent concrete aggregate. That was a main use of the rock, plus ballast for the railroad we eventually built.

HKS: How many engineers were there? Who had this kind of experience? You built roads in Pennsylvania. You had some sense of at least the basic physics this would require.

RJG: I was the only engineer, until we got close to building the pulpmill. Then I hired a few others.

HKS: Oh, I thought there were lots of engineers running around.

RJG: No, no, I was the only one.

HKS: O.K. You must have been pretty busy.

RJG: I was busy all the time. All my life I've been accused of trying to do too much [laughs] by myself. But in the process I saved Mr. Ludwig, I think, hundreds of millions of dollars.

HKS: But you felt generally confident you had the technical skills. There are so many specializations in civil engineering, and you had to have them all.

RJG: To put it in perspective, I grew with the project. We started out small, just trying to plant twenty-five thousand acres a year. The first two years we did not succeed. This gave me time to learn. It gave me time to build up my construction crews. I eventually had excellent [emphasis] Brazilian supervisors and superintendents. They were the ones that got the job done. The engineers helped.

HKS: But did you send away for technical manuals? You figured out that laterite was excellent aggregate for concrete. How did you know these things? Experimentation, or were there books on—?

RJG: In the case of the road surfacing material—the laterite and the river run gravel—experimentation. The rock was obviously a good concrete aggregate. Of course, I looked it up in my engineering books, too. Ludwig had geologists in the Amazon, and I would consult them occasionally. I had them look at it and they said that diabase is one of the best, hardest, and densest rocks on the planet. It will be an excellent concrete aggregate. These are the geologists that discovered the kaolin deposit on the project, which I inadvertently told them about. I didn't know what it was. [laughs] I don't know if you heard about the bauxite deposit.

HKS: No.

RJG: An area farther up the Amazon River, this actually wasn't in the same area of the Jari project, turned out to contain one of the largest bauxite deposits in the world. Ludwig never developed it, he eventually sold it.

HKS: Did you actually test the material?

RJG: No, we didn't. When we actually had to produce structural concrete for foundations for the pulpmill, then we did test it. But we already knew it was good stuff, and it tested very well. The initial uses for concrete were for housing, where we weren't concerned about strength.

HKS: Did you feel your isolation, the distance you had from technical support, was a serious problem?

RJG: I considered it an advantage. [both laugh] We were not bothered by extraneous circumstances.

## Forestry Activities

HKS: Tell the forestry story.

RJG: Before I got there in the first dry season, about one thousand acres had been cleared. These one thousand acres were planted from seedlings grown in the original nursery at Monte Dourado by Tom Bunger. The first year I was there, or really the beginning of the second year of the project, about seven or eight thousand acres were cleared, and Tom Bunger did not have enough seedlings to plant this area.

There were two forestry debacles in the early years. That was the first one. The second one was when most of the seedlings that were planted died because of a drought. The reason this guy got fired, like I'm saying, was that he didn't have enough seedlings in that original nursery to plant even the smaller amount of land that was cleared. He was a young guy. I was twenty-nine at the time, and I think he was about the same age as me.

HKS: It seems rather fundamental, from a distance, he needed enough seedlings per acre and so forth.

RJG: Yes.

HKS: Why didn't he?

RJG: This is going back close to thirty years. I just can't remember the details about why there weren't enough seedlings.

HKS: But in Ludwig's plan, if you made a mistake like that, you weren't the right guy for the job. Is that right?

RJG: This was a guy fired for an obvious technical blunder, yes. Now, after this forester was fired, we went for a long time with no forester on the job. Ludwig hired a construction guy to run the project after this. At this time the general manager living in Belém was eliminated and the general manager had to live at the site.

HKS: Clayton said when he was almost fired his first year, he went to somebody in Caracas, an engineer type. Clayton had predicted that the seedlings would die.

RJG: Yes. Clayton definitely predicted that, and Clayton is the one that told Ludwig what was happening, and prevented further disaster in doing so.

HKS: I ask this delicately. Were there any engineering debacles? I mean, the forestry ones got a lot of press. People wanted Jari to fail, I guess, because all we read about in the States was, it's bad forestry, it ain't working. Were there engineering problems?

RJG: I made a few mistakes over the years. But they were relatively minor.

RJG: So you were essentially chief engineer on site during all your time there.

RJG: I was the only engineer at first. I was really the manager of engineering and construction for the project. It was a division like forestry was a division, and we reported to the general manager. One mistake I made which cost Ludwig some money was the original water system, potable water for the town of Monte Dourado. Instead of using water from the river and treating it, I thought wells would be better. I had wells drilled and pumps bought for them. It turned out to be very difficult to pump the water, and the pumps were made in Brazil, and they were no good. The well system did not work. We eventually built a water treatment plant to use river water. I think that's probably my worst single engineering error in the years I was there.

HKS: But you obviously didn't get fired for it.

RJG: I don't think Ludwig ever realized [both laugh] this particular problem.

HKS: When did the foresters arrive? My perceptions are what Clayton told me, and he came a little later in the game.

RJG: Yes.

HKS: He said that you guys were scraping all the topsoil off with your damn tractors.

RJG: Yes, but we were doing it with the forester there. This was part of the problem of the first planting. Ludwig never conceived of clearing by hand. To Ludwig the way to clear was to get the biggest bulldozers you could, which would get the most number of acres cleared per hour, per day, however you wanted to measure it. This is the way the clearing started, and the roughly eight thousand acres cleared the first year were all done with large bulldozers, Cat D-8s and D-9s. They were fitted with special clearing blades called KG blades. They were made by the Rome Plow Company in Georgia. The things were designed so that you would get a minimum of soil disturbance. The blades were designed to shear the trees off at ground level, not uproot them. Well, that worked well on your softer trees, but the extremely dense Amazon hardwoods, I mean the thing just wouldn't shear. They eventually had to be uprooted to get them out of the way. So you did get soil disturbance in spite of having the special blade.

What was cleared was pushed into windrows, and then the windrows were burned. So even if you had no soil disturbance, you would obviously have more nutrients where you burned in the windrows than between the windrows. With the soil disturbance, you had the little bit of topsoil plus the nutrients from the burning in where the pile was, and not much in between.

So that when the area was planted, you get dramatic differences in the rate of growth. The gmelina shot up where the windrows were, and practically didn't grow at all between the windrows.

HKS: How many tractors did you have? Ten? Fifty?

RJG: When I got there we had nine D-9s, and we then got twenty-one D-8s. Whatever we could get with those nine D-9s originally was what we were supposed to clear, and I forget what it was. The first year we were supposed to get twenty thousand, twenty-five thousand acres was when we got the twenty-one D-8s, which was in the dry season of '68, six months or so after I got there.

HKS: So that part of your job would be making sure spare parts came in, and mechanics. The maintenance of that equipment in that location must have been a major challenge.

RJG: This was not only a major challenge, but the most difficult of all the problems, right.

HKS: How about the mechanics? Brazilian mechanics?

RJG: Brazilian mechanics are very [emphasis] good, if properly supervised. I have to back up. When I got there, the general manager had hired the Brazilian subsidiary of Morrison-Knudson to actually run the project and do the clearing. There was an American superintendent on site at the time, a guy named Bob Romans. He did have an engineer there, that's right, when I got there. I'm forgetting this. But his job was essentially to measure the areas cleared, not to build any roads yet. At any rate, they only lasted a few months. I mean, three or four months after I got there, Ludwig threw the general manager and them out. They were really not significant players, let's say, in the project. And then the other engineer left, and that was it. I was the only one for many years like I've already explained.

HKS: But your basic supply of spare parts; was there a major stockpile in Brazil itself?

RJG: You had the local Caterpillar dealer in Belém, who was supposed to have a stock of spare parts, but they never were able to meet our needs. This project was probably ten times their total volume per year, previous to this project starting up. Essentially the parts had to be ordered out of the United States.

I was never responsible for maintenance of equipment. Maintenance was also a separate division, and this created problems. Maintenance had to maintain the forestry equipment and the construction equipment. Of course, initially it was all construction equipment. It was a separate division, but I certainly got involved in it, because I needed to keep my machines running.

HKS: Did you have a lot of problems?

RJG: Once we got the logistics set up, it was not that great a problem. With the riverboats that we had, we had the means of supplying the project with one or two boats a week from Belém. And you could air freight. Belém has an international airport. You could airfreight emergency parts into Belém, get it on a boat, and get it out there reasonably fast.

HKS: A D-9's pretty big. Did that come in on a barge?

RJG: They were put on barges, flat top barges, the initial one.

HKS: There had to be port facilities when you got there obviously to off-load. I don't know what a D-9 weighs.

RJG: Forty-five tons.

HKS: Forty-five tons.

RJG: It's not difficult to off-load from a flat top barge. Just cut down a few trees and pile them against the bank and run the barge ashore.

HKS: Oh, I see what you're saying. Just drive it off.

RJG: Fall a few trees and get them positioned so you can walk it up. It's not difficult to off-load heavy equipment, at least Caterpillar tractors off a flat top barge.

HKS: Back to forestry. Was Clayton's arrival significant? He came down as a geneticist to work in the nursery. He wasn't really in a decision making position officially when he came down. I don't know if I'm getting ahead of the story or not, but when forestry's sort of taking off on its own.

RJG: O.K., we've gone over the first forester and what happened to him. Then we have the interval between him leaving and Clayton arriving. That interval was something like a year.

When I got there, February of '68, it was already the beginning of the rainy season. The rainy season essentially starts in January. That original nursery, the one located in Monte Dourado and planted by this forester who got fired, was to plant what was cleared before I got there. It was only about a thousand acres. This was planted in the rainy season of '68 under this forester. O.K., he didn't get fired then. He had enough for those thousand acres. It was the next year. The next year is when we cleared about eight thousand acres. That's when they ran out of seedlings. They didn't have enough to plant those seven to eight thousand acres.

HKS: O.K.

RJG: That would have been the '69 rainy season. Tom Bunger, the forester, left. Then Clayton got there in time to establish the nursery for the 1970 planting season. There was probably only a few months between Tom Bunger leaving and Clayton arriving. I know they never met. So Clayton gets there in the second half of '69, and he has to establish a nursery to plant twenty or twenty-five thousand acres, which now we were clearing by hand. We were going to meet that goal for the first time. It was decided to give up machine clearing and go to hand clearing, just falling the trees in place and burning the trees in place. This decision was made between the '69 and the '70 planting. Clayton got there with the responsibility of establishing the nursery and having the seedlings to plant twenty-five thousand acres.

O.K., so we built the nursery. Clayton got the seedlings planted for twenty-five thousand acres, and we got that amount cleared to be planted in 1970. This, then, is when the seedlings died from drought. The rainy season delayed in coming in. We had a couple of rains early. They started planting, then it stopped raining, and a very large percentage of the seedlings died. You'd walk out there and watch where they planted, and you'd see them wilted. [laughs]

HKS: Did you help construct the nursery? Was that part of your job?

RJG: There was an elaborate storm drain system to keep the nursery from washing away in the intense rain. It was a sandy soil area. Erosion control was a major engineering problem for the nursery and really for the entire project.

HKS: By then he had a corner or someplace, he was growing pine seedlings, too. Somewhere along there he gets—

RJG: Right. Somewhere in there he started experimenting with pine, and he probably told you the story of how he got Ludwig to see the pine.

HKS: Yes, but you tell it.

RJG: He got Ludwig to see the pine by driving him past the pine, never saying a word to Ludwig, which is the way to do it. Ludwig said, what's that? And Clayton said, that's pine. He said, Stop here! Hey! That's good looking stuff! [both laugh] Because if he had asked permission to do it, he probably would have gotten fired.

HKS: Sure.

RJG: Ludwig was set on gmelina.

HKS: What were Brazil's concerns about the project? Was that a constraint on you, that the public or the government of Brazil. Was there some sensitivity to this at certain points of the operation.

RJG: There was no real problem in the first years. When I got there, the president of the company was a Brazilian general, an ex-general, General Tubino. At that time the military ruled in Brazil. In fact the years that the military ruled in Brazil is when Brazil made the most progress that they've ever made in their history. Around June of 1968 another military man, an ex-captain, was hired—Heitor Ferreira—who eventually became the secretary to the president of Brazil that succeeded the one that was in office at that time. Heitor Ferreira was very capable. Ludwig liked him a lot, and reluctantly let him resign from the company because he knew he was in line to become the secretary of the next president of Brazil.

To the Brazilian public, the project was always looked on, right from the beginning, as a cover up, because it was difficult if not impossible for the average Brazilian, or any Brazilian probably, to believe that we were in the Amazon planting trees. [both laugh] The fact that we were spending a billion dollars doing it didn't seem to penetrate. Here we were, wiping out the native forest—and that's just what we were doing—wiping it out and burning it up. The only wood we used from the native forest was what we used for construction, which was of course a tiny percentage of the actual forest cleared. You'd talk to the average Brazilian on the street, and he would say, Come on, you guys aren't out there planting trees. It's a cover up for your gold and diamond mines, right? But this never hurt the project in the initial years. It only hurt the project as the project grew. About the time we were building the pulpmill the project got so big that it got all this bad press from the Brazilian press. Which of course is what eventually forced Ludwig to give the project away.

The reason that the project was taken from Ludwig, stolen from him by the Brazilians—the reason he lost six hundred million dollars of his own money—is because he did not give value to Brazilians. He never realized, in spite of me [emphasis]—and I was the only one that tried to point this out to him—that the talent existed in Brazil to do everything he wanted to do. You did not have to go outside the country. He hurt himself, you know, by hiring so many foreigners, outside of Brazil, and by giving zero attention to public relations. If we had this to do over again; Ludwig or his successor would still own the project. We would have built the second pulpmill. We would have built the papermill. We would have built the hydroproject. I'd still be there building, hopefully, which [laughs]—if he had only paid attention to this detail. But, you know, it's sad, but it is the main reason for Ludwig's demise in Brazil.

HKS: Give me an example of how you would recruit certain skills. You'd run an ad in the paper in Rio or something?

RJG: Right. That was one way to do it. But most of the talent existed in northern Brazil, or at least northeast Brazil. Most of your skilled labor in Brazil are people of modest origins. So the people working for these big construction companies on huge development projects all over the country tended to be from northeast and northern Brazil. They migrated to southern Brazil, where they were hired and trained by the large Brazilian construction companies.

HKS: I see.

RJG: You might catch them even in Belém. We recruited a significant number of people. But we advertised in papers all over the country and got really qualified people.

HKS: Clayton was saying that it was difficult to recruit technical people from southern Brazil. It's like going to Oklahoma if you live in New York.

RJG: Yes.

HKS: The Amazon wasn't the place you wanted to be—

RJG: And he's right. Highly [emphasis] technical people, like foresters and engineers. But I needed relatively few engineers. We were mainly construction. We didn't design the pulpmill. The pulpmill was designed by someone else. We built all the onshore facilities for the pulpmill. It was about seventy million dollars of the three hundred fifty million total that the pulpmill cost. All that was built by us on site.

HKS: With all the shifts in managers, how did that affect your ability to cope from day to day? They went through thirty directors in ten years. There must be a lot of confusion.

RJG: [laughs] There was. I just happened to be the type that could adapt. Eventually most of the general managers were superfluous. I mean, you would treat them politely as your boss, but your real boss was Ludwig. Now, there were a couple of general managers there that were good and that were trusted by Ludwig, not the least of which was Elmer Hahn, who was there during the construction of the pulpmill. Mr. Hahn and Mr. Ludwig were like that [gestures], and talking to Mr. Hahn was almost like talking to Ludwig. But he was the exception.

HKS: But did you waste time, get started in one direction and the new guy would come in and say, No, that's wrong? Or that didn't affect what you were doing? Did it affect what you did, with all these turnovers in general managers?

RJG: It didn't affect me because I was somewhat of a tyrant in this respect and a bit ruthless. This project was my whole life, and I wanted it to succeed. I didn't let these things deter me from designing and building what I thought the project needed. In a nutshell, that's it.

HKS: O.K., it's hard to imagine that there wasn't some inefficiency from all this turnover in general managers.

RJG: To this date there's probably still some doubt as to the viability of the Jari project as a forestry project. I can't say whether it was successful or not. I think as a construction project it was definitely highly successful. We got what we needed to get done, on time, and at not only a reasonable cost but an amazingly low cost. I don't know if you realize that the estimated cost of building the pulpmill on site by traditional means was about four hundred fifty million dollars. And by doing it the way we did it, by building the actual mill in a shipyard in Japan and towing to the site, we cut about two years off the construction time and saved about a hundred fifty million dollars. It was built for about three hundred million.

HKS: And even to Ludwig a hundred and fifty million dollars is important, right?

RJG: Right, exactly. The only thing, the only thing that he did not finance himself, by the way, was the pulpmill. He got financing for that, three hundred million.

HKS: Obviously you needed construction camps, but they were temporary, right? This is where the workers lived in order to build what they're building. Or did those camps turn into towns?

RJG: They turned into towns. The town of Monte Dourado would not exist, I guess, if it weren't for me. No, I don't guess. It wouldn't exist. The project had grown so much that we, in order to retain supervisors and even foremen over the long haul; it wasn't just a two-year construction project. This was a ten-year construction project or more. These people needed to have their families with them. O.K., it was also obvious that, over the life of the project, the people that were going to operate it needed permanent housing. So, O.K., let's start building the housing now, and we'll house the construction people for this very long construction project, which then will eventually be turned over to the operating people. So I started building houses in Monte Dourado around as early as 1970. Kind of sneaked the first ones in there, at least for this type of personnel, not for staff, your expatriates.

RJG: Did you saw your own lumber for that?

RJG: Yes, we sawed our own lumber for that.

HKS: And that was adequate for your building needs.

## Housing Situation

RJG: For the first couple of years. And then, Mr. Ludwig also realized he needed to build housing. This was somewhere in '72-'73. He realized this too, that we need to start building housing for our permanent staff, but our construction personnel will live in it temporarily. So he then sent to us a set of concrete forms to build concrete houses. We started with wood houses, right, and we were sawing our own wood. At that time he owned the company that made these concrete forms. They were aluminum panels. He sent them to us, and it turned out to be a very good type of housing for the tropics and for what we were doing. In spite of having this rot resistant wood, the concrete was better yet. We had the rock and the sand for the concrete, which is ninety percent of the volume of concrete. All we had to do was import the cement, bring it in from outside.

HKS: Interesting. Interesting.

RJG: It took off. Monte Dourado went from a construction camp to a town of twelve thousand people by the time I left in 1980, by the time the pulpmill was built. We also built satellite communities for the plantation, and plantation management, and harvesting—three satellite communities away from Monte Dourado, each one of about three thousand people.

HKS: Was part of the disaffection of the Brazilian press and public that you had inadequate schools, hospitals, and so forth? Or was it just that they didn't like the idea you were there at all?

RJG: It was strictly jealousy that Americans, gringos, were in their precious Amazon and developing it, which is something they had not succeeded in doing in the three hundred fifty years they owned it. Period.

HKS: So some of the opposition could have been high up in the government?

RJG: The government people that knew of the project and understood it, were for it. But the bad press, the pressure became so great against the project that even these people were afraid to speak up in its defense. This hit its peak about the time I left, around 1980, after the pulpmill was up and running and starting to export and get some money back.

HKS: We've talked about building the nurseries. You said that Monte Dourado was in the wrong location, as it were. That's because it grew? It wasn't designed to be a full-fledged town originally.

RJG: No, it's a town, it's a full-fledged town with all the amenities: supermarket, school, hospital, slaughterhouse, and you name it. It's just that it's neither near the plantations nor the pulpmill. There's sixteen kilometers between the pulpmill and Monte Dourado. So everyone that works at the pulpmill has to drive that sixteen kilometers every day.

HKS: And the pulpmill was where it was because—

RJG: The pulpmill had to be where it was. It's essentially the head of navigation on the Jari River. Upstream from the pulpmill the depth of the channel goes from an average of ten meters to something like three meters.

HKS: I see.

RJG: There was no choice. The pulpmill had to be where it is to get ocean going vessels to it. We had to do some dredging, but it was relatively minor, compared to what would have had to have been done if we had put it farther upstream.

HKS: Clayton tells a story that delighted him. Crown-Zellerbach was sending a ship, and you hadn't built a pier yet. You built a pier in two weeks. That may not have been you. I assumed it was.

RJG: Oh, yes.

HKS: And he said that was standard procedure. If you didn't need this decision making process, you had to build a pier, and you build it.

RJG: Things like this were a dime a dozen. They were always doing something like that. To build a simple wooden dock to tie up an ocean going ship to, that's easy. You don't need these elaborate docks [both laugh] to dock them.

HKS: There's something in some of the articles I read. A town grew up across the river, and that's where all the gambling was, and so forth. Was there a pure town on one side and a corrupt town on the other, or what?

RJG: This is inevitable in Latin America, and probably any Third World country. Whenever you have any kind of development, money coming into an area, you get satellite communities. It attracts people, and it seems to be impossible to prevent it. This was a slum that grew up across from Monte Dourado. It was in a swamp. We had high ground on the Monte Dourado side, and there's swamp on the other side. And this town was on stilts. The population of this town more or less grew with Monte Dourado. I never made an actual count, but they probably had twelve, fifteen thousand people there when we had twelve thousand people in Monte Dourado.

HKS: And what was the name of that town?

RJG: They call it the Beiradao. A translation of this is very difficult, but it denotes—it's typical of the Amazon, and there's probably nowhere else on earth like it. And it speaks to the size of the Amazon River and the immensity of the river's edge. The river's edge is called the *beira*, the edge of the river, *beira de rio*. And when you add d-a-o to it, you make it larger yet, immense.

HKS: I see.

RJG: Immense. The immensity of the endless edges of the rivers of the Amazon. This is really what it means.

HKS: So other than gambling and whatever, were there stores there where you could buy anything; you could buy a refrigerator across the river.

RJG: Yes, right. They grew up, so you could buy a refrigerator—

HKS: They'd sell you, plus markup. [laughs]

RJG: —buy anything you wanted. [both laugh]

HKS: Interesting. You talk about main roads and the railroad. Some journalist made a big deal out of Ludwig's obsession with the railroad, like he was—like he was at fault for having a railroad. So what do you think happened on the railroad. What's your story?

## The Jari Railroad

RJG: The railroad. This was of course one of my major construction projects. The railroad was only feasible if the plantation went to five hundred thousand acres, which was Ludwig's ultimate plan. Five hundred thousand acres and twelve thousand tons a day of wood to get to the mill was unfeasible trucking it. It all had to come on one road to the port, at least for a portion of it, right? About twenty kilometers. So when Ludwig first mentioned a railroad, I laughed. [both laugh] A railroad [incredulously], the way we're spread out like this? And I stopped laughing. I mean, he was dead serious about it. We're going to build a railroad, and—

HKS: So it was his idea originally that—

RJG: Yes.

HKS: Based upon this gut feeling he had?

RJG: Yes.

HKS: Why would he feel that way?

RJG: Just from what I'm telling you. He conceived things like this. Remember, Mr. Ludwig only had an eighth-grade education, but he could definitely see things that were obvious and that were practical, that made sense. He was intent on a five hundred thousand acre plantation, started at two fifty, but somewhere in there he jumped it up to five hundred. Then I had to agree. It didn't matter if I agreed or not. But I did agree. Five hundred thousand acres and then that volume. You had to move it with something other than trucks. In spite of our surfaced roads and all—and we kept them open through the rainy season by constantly blading them. We hauled on these roads for the worst times of the rainy season. Remember, we got a hundred inches a year of rain there. But those volumes really justified a railroad.

HKS: What did you know about building a railroad?

RJG: Building a railroad is not much different than building a road. We hired a consultant to give us the basic parameters, an outfit called T. K. Dyer near Boston. They were rail experts. Minimum radius of curvature for the railroad, superelevation, maximum adverse grade, maximum favorable grade, and then they designated the rolling stock, the size of the rail cars and the size of the locomotives, etc. We took it from there. We did the surveying and laid out the railroad in accordance with their parameters. It's not much different than building a road. To actually lay the track, I hired a guy—he happened to be an American from a manganese project just up the Amazon River from us. It was a Brazilian company, but they had retained this guy. He built the original railroad there for this manganese project, a couple of hundred kilometers of railroad.

HKS: This was Antunes.

RJG: It was Antunes, right, owned by Antunes. Ludwig's, quote good friend, unquote. (Antunes turned out to be anything but a good friend.)

I borrowed this guy from the Antunes project, and he ended up charging Ludwig a hell of a price for him. This guy organized the track laying crew and showed them how to lay track and did it. Now we did, like I said, all the grading and the drainage structures, and everything else on the railroad we did ourselves. It turned out to be an immense earthmoving project. The pulpwood had to come from the north and from the west to Munguba (name of the place where the pulpmill was built). Going to the west, the streams flowed north and south, so you were always going cross drainage. Then going to the north the streams flowed from west to east, so you were going cross drainage there, which lent itself to huge cuts and fills, because of the low grade requirements on the railroad. One and a half percent was the maximum favorable grade.

HKS: I understand why you choose the size of a locomotive. I thought railroad cars were standard size. Are they custom made, longer or shorter?

RJG: We were not allowed to use U.S. standard gauge. There are two standard gauges in Brazil, meter or narrow gauge, and meter sixty. We had to use meter sixty, standard Brazilian gauge. So they were built specially to begin with. And the railroad cars to haul the pulpwood were specially designed, as I remember.

HKS: I see.

RJG: There was nothing very unusual about them. The length of them was determined by how long the average or the longest piece of gmelina was going to be, this sort of thing.

HKS: Two main branches, then?

RJG: Right. One to the north and one to the west, right. We built forty-four kilometers of railroad.

HKS: So the locomotive goes in reverse in one direction pushing the empty cars.

RJG: Right. Diesel locomotives, they don't even know whether they're in forward or reverse. They don't care.

HKS: The maintenance for a diesel locomotive was no problem. You had access to engineers and spare parts. That sort of logistics had been worked out.

RJG: The railroad maintenance was relatively minor, you know, compared to everything else that had to be maintained at the pulpmill.

## Airport and Kaolin Deposit

HKS: Is there something equivalent to the FAA in Brazil? You had to have certain standards for an airport?

RJG: The airport grew kind of with the project. Yes, to operate first the DC-3 and then, we eventually got an F-27, twin engine turbo-prop which carried forty to fifty passengers. There were minimum requirements for the length of the runway and airport to handle this type of aircraft, which we exceeded, I know. We did not pave the runway. It was, again, a laterite surface runway. And we had a huge hangar.

HKS: How far was the airport from the town?

RJG: The airport was about five, six kilometers from the town. There's a flat plateau about two hundred meters high above the river, a few kilometers from Monte Dourado, which happened to be the first area that was planted to gmelina. And the airport was up there. It is an excellent location for an airport. I remember we could have gotten a runway five thousand meters long if we needed it in that particular location. As it is, I guess the one we built was about two thousand meters long.

HKS: Most of the air traffic was with building?

RJG: Up to the time I left there was no commercial air traffic. Since I left, there are commercial flights into Monte Dourado. Everything was company plane up till the time I left.

HKS: But most of the flight was the relatively short flight down to Belém.

RJG: Three hundred miles.

HKS: The kaolin processing plant. You didn't know what kaolin was, but you described it as, what, slippery stuff? Or what is it?

RJG: The first place cleared is where the airport was, this flat plateau above Monte Dourado. The only stretch of road that was built when I got there, fifty feet or so below the top of the plateau in this road cut there was this very white clay. I wondered what it was. I thought it was unusual, but I didn't recognize it as anything valuable. Anyway, I mentioned it to someone. I forget the details of this, but eventually one of Ludwig's geologists came out to look at it. He looked and said, That's kaolin. I said, What's kaolin? He said, Well, high grade kaolin is used in making paper. It's used to coat paper. Highest grade paper is coated with kaolin. He said, There's considerable use for it. This geologist then started looking at the size of this particular deposit and at other plateaus in the area. Across the river, to the east of Monte Dourado, in one plateau he discovered what turned out to be at the time the largest kaolin deposit in the world. This was a sideline to the forestry project. We developed this deposit. We opened up a mine over there and built a slurry pipeline from the mine across under the Jari River to the site of the pulpmill, which was also the site of the kaolin processing plant. The kaolin processing plant was built before the pulpmill, and we built the kaolin processing plant.

HKS: Then you shipped that out by barge, out to the world market?

RJG: Not barge but boat, ocean going ship.

HKS: Did the Brazilian paper industry use it?

RJG: No, it was almost all exported. Now for the construction of the kaolin processing plant, we did all the site preparation and the civil work, but this is one of the few things that was contracted out. But the mechanical, electrical, structural steel, etc., was done by a major Brazilian contractor. The kaolin processing plant was built on site from the ground up, as opposed to the pulpmill, which, as you know, was built in the shipyard in Japan and towed to the site.

## Owning the Land

HKS: Clayton talked a little bit about, and the press talks a little bit about the actual ownership of the property in First World terms, what Ludwig bought and what he actually owned. There were title conflicts and things like that.

RJG: Yeah:

HKS: Did that affect your job at all? A very valuable mineral deposit, would that ever have been a challenge to Ludwig's ownership?

RJG: I don't know if Ludwig realized what he was buying. The company that owned this property presented it to him as a solid block. They were politically powerful people that owned this company, especially in the State of Pará. But in reality it was a hodgepodge of individual titles with poor or nonexisting descriptions of the property boundaries. So that when you tried to plot this on a map, you had a bunch of overlapping properties along the rivers and you had a huge hole in the middle, because the properties didn't extend to the middle. This didn't bother anyone in the first years. It only became a problem when the project started attracting all this attention and the bad press. I mean the Brazilians jumped on this—he doesn't really own this land. Well, no one in the Amazon does. If you own along the river, you own what's inland from it, essentially, in the Amazon.

HKS: Did Ludwig pay royalties or taxes on the value or...?

RJG: That's a good question. I don't know.

HKS: I mean, there's a manganese mine, there's a bauxite mine, and...

RJG: Right. I never got involved in that. I don't really know if he paid royalties on the kaolin, for example. He never did develop the bauxite. Property taxes at that time in the Amazon were nonexistent. Since then, property taxes have become significant for anyone who owns land in the Amazon. For example, I owned nine hundred hectares of Amazon land, which I just sold, and, like all good Amazon landowners, I have not paid my taxes. But I owe a considerable amount. I sold it to this individual, he realizes that the back taxes are owed on it.

HKS: I can see the public image suddenly in Rio. There's this billionaire not paying any taxes or any royalties.

RJG: Right.

## Construction Projects

HKS: I have three more items under construction projects. I don't know the sequence. Do you want to leave the pulpmill until last? You also have the power line and the sawmill that's on, in there.

RJG: They're all together. The power line was just a minor project. I think it was a thirty-three kilovolt or sixty-six kilovolt line from the pulpmill to Monte Dourado, sixteen kilometers. It was no big deal.

HKS: Which way was the power going? To the mill?

RJG: No, it was going from the mill to Monte Dourado. You see, Monte Dourado was supplied with electric power by diesel generators until the pulpmill was built. And the pulpmill, which had a fifty-five megawatt generator, had enough power to supply Monte Dourado. We required maybe three megawatts at Monte Dourado. So by building this power line we were then able to shut down the diesel generators, which had been running for twelve years or so.

HKS: And diesel's pretty expensive.

RJG: We burned about twelve thousand liters a day of diesel fuel to supply power to Monte Dourado. Three thousand gallons, and it was very expensive, right. Caterpillar diesels, they're big ones, twelve and sixteen cylinder engines.

HKS: Another reason why it would have been better to have the town closer to the pulpmill.

RJG: Yes. Right. Another good reason.

HKS: And the sawmill was—

RJG: The sawmill was built along with the pulpmill. While it was a major project for us, it was part of that roughly seventy million dollars worth of onshore facilities that were built—onshore as opposed to what was built in the shipyard in Japan. That consisted of your wood handling facilities, all your conveyor belts, etc., the wood yard, and warehouse, and office building, chemical plants, and the sawmill. I mean, the sawmill was significant, but as a construction project it doesn't have to be separated from the pulpmill itself. It was right next to it. We did the site preparation, the concrete work. There was a Brazilian contractor, a major Brazilian construction company at the pulpmill, too, that did all the structural—again, structural steel, mechanical, electrical work. We did not do that.

HKS: But did you actually export any lumber, or was this for local construction primarily?

RJG: No. At the time it was the largest hardwood sawmill ever built in the tropics. The lumber was definitely for export.

HKS: Is that right? I didn't realize that.

RJG: Yes. And the objective was to export. This started well after the pulpmill and after I left.

HKS: Did you deal directly with Brazilian officials routinely on your job?

RJG: No, no one was inspecting us. I dealt considerably with the navy, because anything that you do along a navigable waterway must be approved by the navy. And we were sure that we got that approval.

HKS: The navy had expertise on its payroll, rather than sailors?

RJG: They had a couple of guys that weren't too bad. They had to approve. We'd submit the permanent dock we built for the pulpmill. Temporary dock you didn't have to worry about. You'd just tell them about it. It's a temporary dock. It's going to be timber piling and then we'll remove it. The permanent dock you had to get approval for. This is where we're going to ship our pulp from. Navigation buoys in the Jari River, in the stretches that we dredged, we had to have the buoys approved, and their location approved.

HKS: In the States the Corps of Engineers provide the same kind of function.

RJG: Right. Exactly. Then the same thing for the airport and our radio that had to be approved by the equivalent of the FAA. They were bugging us all the time. There were many complaints because of speaking English on the radio.

HKS: Was this pretty simple? Was it slow? Where there a lot of delays, fiddling around? Or was it efficient in getting inspected and located?

RJG: Oh, you had to struggle and wait, being a foreign company. Sometimes they were O.K., sometimes they'd create obstacles that really didn't need to be created.

HKS: You find that in the States.

RJG: You find that in the States. I had relatively little trouble over the years. It was just those last years, when that wave of bad publicity just overwhelmed the project. We could do nothing right. I went to a meeting in Brasília with the minister of energy, at the offices of the Ministry of Energy. At that time we were trying to get approval to build the hydroproject. I went prepared, had all our studies, everything we'd done. They couldn't have cared less. I mean, their minds were closed to even listening, you know, to what you had to say at that point. This was in 1980.

## The Pulpmill

HKS: Let's talk about the pulpmill. That's one of the key figures in the whole story. At least it's portrayed that way.

RJG: Like I explained earlier, I don't think the decision was made to actually build a pulpmill until '74 or '75.

HKS: So the assumption was just shipping pulpwood by barge, how would you get it out of there? That was going out to the world market—that was actually the purpose of the forestry project initially, as far as you can determine.

RJG: No. It was never [emphasis] just pulp. It was pulp, lumber, and plywood initially. Because of the rate at which gmelina was expected to grow. Ludwig was envisioning at that time, at the beginning, gmelina that big [gestures] in diameter after ten years.

HKS: Twenty inches or something.

RJG: Yes. Some of which would be suitable for peeler logs.

HKS: I didn't realize that. I thought it was only pulp.

RJG: No. Then after it became obvious that the gmelina wasn't going to grow to that diameter in ten years, no one ever talked about plywood again. There was still talk of saw timber, lumber, but gradually the shift went to pulp and paper, high grade paper. I guess gmelina apparently tested very well for making high grade paper. Here I'm talking in a field I don't know anything about, but this is my understanding. So that eventually the industrialization of the project turned entirely to pulp and high grade paper, with the sawmill to saw the tropical hardwoods that would be cut to expand the plantation and to supply fuelwood to the generator of the pulpmill.

HKS: The sawmill we're talking about, this may not be the one that Clayton was talking about, but that was one of the few battles he lost. It was a huge sawmill. It was never efficient. It may not be the same mill we're talking about.

RJG: No, I'm pretty sure it is. It's the hardwood sawmill built at the pulpmill.

HKS: Cutting, obviously, the local logs.

RJG: Right. Unfortunately I left shortly after it went into operation. I watched them saw a few times when they were just starting it up. I don't have any information on how that turned out.

HKS: So the pulpmill, obviously, is a defined thing. You need the labor force, transportation systems—

RJG: Oh, listen, let me give you the economics of the Jari project in a form that probably no one has ever given it to you in.

HKS: O.K.

RJG: When this scenario developed I'm not sure, but it was definitely in the later years and while the pulpmill was under construction. The project was only feasible, essentially, with two pulpmills; the first one that was built and another one next to it. These two pulpmills would supply a high grade papermill which would produce an income of six hundred thousand dollars a day. Only then would all expenses be covered and the vast infrastructure you had to maintain be covered and the project be viable. Six hundred thousand dollars a day is what had to be generated in income. The pulpmill, and I'm not sure if that involved a second pulpmill or not, because too many years have gone by. It may not have. It may have been just the existing pulpmill plus the high grade papermill next to it. Now that involved another billion dollar investment. It was four hundred million roughly for the papermill and six hundred million for the hydroproject necessary to supply the approximately three hundred megawatts that the papermill would consume. It was another billion dollar investment to get to that point.

HKS: At one point this was on the drawing boards, and, presumably, Ludwig would continue to invest, and add the second mill and hydropower and all the rest?

RJG: Definitely. That was the ultimate project. That was his goal. What killed the project was the government, the Brazilian government, not allowing him to build the hydroproject, because of this bad publicity and the general feeling in the country of Brazil against the project. Without the hydroproject, he couldn't put in the papermill, and without the papermill the project was not viable. So he was faced with operating the existing pulpmill at a loss forever, or giving the project away. And so he gave it away.

HKS: So the hydropower primarily was to run the mill. There wasn't a local market for electricity then.

RJG: Exclusively for the papermill. Apparently these things require vast amounts of power. I've never even been in one myself. We did all the preliminary work, in conjunction with Alcoa, for the hydroproject. We knew the potential that was there and even the size of the reservoir had been determined and the location of the dam, etc.

HKS: When you make paper you have to dry it. I imagine that's where most of your energy goes, the drying of the paper.

RJG: Probably.

HKS: Talk about the pulpmill. Obviously, to the amateur engineer like myself, you want to make sure the foundation is the same size as the thing that floats in, so you put it on there and it matches up.

RJG: O.K. The pulpmill was designed and built probably like no other on earth before or since. The concept of the pulpmill was not only for this pulpmill. The concept that Ludwig came up with was really for a method to build efficiently and cheaply in remote locations in developing countries. By building the pulpmill in a shipyard in Japan and towing it to the site, you eliminated having to have a very highly skilled construction work force on site, and the logistical problem of getting all those individual materials to the site. Ludwig envisioned this method of construction being used indefinitely.

HKS: I see.

RJG: Saudi Arabia needs a desalinization plant in the remotest corner of their kingdom, or maybe I should use a country like Afghanistan. Rather than try and build it on site, you build it in a shipyard in Japan or wherever, and you tow it to the site and plug it in and start it up.

HKS: How long was it at sea?

RJG: It was at sea about two months. They were called platforms, industrial platforms, to dispel the image of it being a pulpmill on a barge. The entire structure was a pulpmill, and it was designed to float. But it was a pulpmill, it wasn't anything else. Then when you got it to the site and put it on the pile foundations, you cut all kinds of holes in it to run your pipes and wiring and whatever else you needed to run into it.

HKS: Of course, otherwise you'd have a tough time unloading it from a barge.

RJG: Exactly. Do you understand how it was done, how it was put in place? First, there were two platforms that made up the so-called pulpmill. One was the pulpmill itself and the other was the power generation unit and, I guess, the recovery boiler. Each of these platforms weighed about forty thousand tons. They were eight or nine hundred feet long, a hundred fifty wide or so. In other words, the size of good large ocean going ships. They were built in the shipyard in Japan and floated and towed around Cape Horn, thirteen thousand miles or so. Each one was pulled by a twenty-five thousand horsepower tug. The main business of these tugs was towing offshore oil drilling platforms, for oil companies around the world. That's why these tugs even existed. While the mills were being built in Japan, we were preparing the foundations for them on site. The foundations for each platform were about eighteen hundred wood piles of a very rot resistant Amazon hardwood.

This was obviously a major pile driving project, eighteen hundred piles for each platform and the average length of the piles was—I keep mixing metric and English systems. Let me keep it in metric. It was about fifteen meters, thirteen to fifteen meters. The longest ones were about thirty meters. They had to be spliced. The platforms were designed simply to be floated over these piles and sit on them. There was no physical connection. They were not bolted to the piles or anything like that.

HKS: So you float them in at high water time?

RJG: I'll get to that.

HKS: O.K.

RJG: How to float them over. All we did was, we had to line the piles up. We had a very elaborate location drawing of the piles, because they had to set under the structural members of the platform, obviously.

HKS: Sure.

RJG: And we followed this. And the piles had to be cut off at exactly the same elevation, with about a plus or a minus three millimeter tolerance. Then on top of each pile was—a two inch thick neoprene pad was placed, which was the actual thing that the steel of the platform sat on. Now, to get the platforms over the piles, we diked the entire area. We built a dike around the entire area where the platforms would sit. To one side of this dike we dredged a channel in from the Jari River. So we had a channel in from the river, and then we had what was called a berthing basin, which was an area wide enough for the two platforms to float, side by side.

The platforms got there and we guided them into the berthing basin. Then we had to close off the river and this channel, and bring that opening up to the height of the dike we had already built around the whole thing, right? Around the berthing basin and the piles. Closing off the river, which was a technically challenging job, to say the least, and all this was done in the rainy season with earthmoving equipment we had there. The dredge has an amazingly large pump, and it's a high volume, low head pump. With pipe from the dredge we put pipe over the dike from the dredge, right? The dredge simply started pumping water from the Jari River into this now completely enclosed basin which includes the piles. We raised the level of the water about seven meters, exactly twenty-three feet in this case, so that the platforms could now float over the piles. We had built substantial wooden bunkers for the platforms to butt against, so that we were sure that they were in exactly the right position.

Floating them over the pilings would have been relatively easy in a calm bay, but we had wind that day, and it was a heck of a job to get them in position. But, very simply, once they were in position against these massive wood bunkers we had built, then all we did was open a forty-eight inch valve that we'd installed under the dike and let the water drain out, and the platforms simply sat down on the piles. No connection whatsoever between the platform and the pile, which is where they sit today.

HKS: So a five hundred year flood would never wash it off, or something?

RJG: No.

HKS: What was the technically most challenging part of that whole thing, other than the sheer magnitude of the job? Getting all the piles cut to the same height, or what?

RJG: No, that was relatively easy. It was done with chainsaws, and we just rigged up a device which would be bolted to the pile to secure the chainsaw, and then you could raise or lower it with a level, a surveyor's level, so that it was in the right position. That turned out to be relatively easy. The river closure was extremely difficult. We couldn't wait till the dry season, that would have meant another six month delay. But it went well—we had stockpiled the right material to close off the river. It was a sandy material, which we could handle relatively easily in spite of the rain.

HKS: How long were the pulpmills in construction in Japan? A year? Two years?

RJG: About a year and a half. We did site preparation in early '76, end of '75, early '76. April of '78 is when we floated the platforms over the piles. The pulpmill went into operation approximately a year later, roughly April '79. From the beginning of site preparation to the end, it's three years and four months.

HKS: And meantime you were—

RJG: It was two years and four months from the beginning of site preparation till we had the platforms in place. After the platforms were in place, the rest of it was hooking up.

HKS: Were any of the engineering problems you faced unique, had never been tried before?

RJG: It may have been the largest wood piling project of all time, for one thing. Each individual operation was probably not unique, building the dike around the piling, and digging the channel in. But the sequence had probably never been done before. All these different operations had never been for this particular purpose of getting a pulpmill in—two forty thousand ton industrial platforms, which were a pulpmill, into place.

HKS: Was Ludwig a help or a hindrance during all this? Was his fertile mind throwing out ideas or was he fascinated by this stage?

RJG: Well, at that time during the construction of the mill, he visited the project about every three months, and he would stay four or five days each time. Was he a help or a hindrance? A help, to me. His imagination was fertile, he came up with any number of ideas, and some had merit and many didn't. He was the type of person that, just because he got the idea, if you could show him it was not a good idea, he wasn't stubborn about it. He'd accept it. He'd say, O.K., you're right. When he did come up with a good idea, and you realized it was a good idea, I mean, you really ran with it.

HKS: What we read about how the Egyptians made pyramids, these engineering feats of primitive societies are rather impressive. You had a lot of labor, a lot of time. Building that mill, putting it on site and having the pilings all lined up, the coordination between you and Japan—probably some of those things that seem hard were pretty simple. I mean, once they had the blueprints to build the pulpmill, you knew where the structural support was going to be. So you knew where the piling had to be, and it didn't come in two feet longer than advertised, or something.

RJG: Exactly. We did our part, and they, the Japanese did their part. One of the most difficult jobs was the actual berthing of the platforms. It was done with our people and our equipment, but the Japanese actually supervised it. Some Japanese came over on the barges just to watch it, and then there were quite a few Japanese on site. A lot of these Japanese didn't speak English, and of course none of them spoke Portuguese, and none of the Brazilians spoke English or Japanese. And [laughs] none of us spoke all three languages. It was very interesting when those platforms arrived, and you had these Japanese seamen shouting to me, giving me orders, and I was supposed to translate, but they couldn't speak English. But anyway, we got through it, because we pretty well knew what we had to do.

HKS: There wasn't the problem that you read about in Japan, where the need to save face is so great—this is a stereotype, now—to save face is so great, that mistakes are allowed to happen, because you don't want to criticize someone. But that wasn't a problem you had?

RJG: No. No. That wasn't.

HKS: Just ordinary communications.

RJG: It was only for that particular operation, because most of your Japanese supervisory personnel did speak English. If I needed to, they could speak English to me, and I could, of course, translate into Portuguese to my supervisors.

HKS: Clayton talked about the Finnish technological help in getting the mill up and running. Were the Finns involved in that early stage right there when you were putting it on the piers?

RJG: No. The Finns only came when it was a few months before it was about to go into operation. The Finns weren't there, and they didn't need to be there for the major part of the construction. I personally had very little contact with the Finns. I mean, I didn't need to, it was between them and the Brazilians that were going to run the mill. And the foresters, of course. Forestry.

HKS: Did you have the road system in, the railroad, everything was ready to go by the time the mill.... They threw the switch on the mill, the wood was coming in?

RJG: Oh, absolutely. Right. I mean that was my major responsibility. I had to have the towns built, the roads and the railroads ready as well as the mill. That was all crucial, of course. We pulled this off, we were able to do it.

HKS: With the kind of financing that Ludwig would manage for this, were you under pressure to not get things ready too soon, because you had capital invested that wasn't generating income?

RJG: No, there was never any worry about getting something done too soon. [laughs] You were always trying to get it done on time. His inception of the project, planting gmelina in the Amazon, I mean logically he would have experimented for ten or fifteen years before actually going into the planting on a large scale. But he was seventy years old at the time. He didn't have time for it.

HKS: Sure.

RJG: So he made assumptions, the assumption being the gmelina would grow. [laughs]

## Leaving Jari

HKS: How soon after the mill was up and running did you leave?

RJG: I left in October of '80. So the mill was in operation a year and a half when I left.

HKS: I see.

RJG: Actually at that point, construction was finished. We had expanded the railroad farther north, and that was our last major project. There were no other major construction projects envisioned at that time. At that time the pressure from the Brazilian press and the public against the project was immense. I made the very difficult decision of leaving, kind of getting out while the getting was good, because I saw what was coming. An opportunity came up with Alcoa. Alcoa was about to invest a billion dollars in an aluminum smelter and a bauxite refinery in the city of São Luís, to the east of Belém. And I hired on with them as a—

HKS: Oh, I didn't realize Alcoa was in Brazil.

RJG: Alcoa's big in Brazil. I went through the Pittsburgh office when I hired and when I left, and that's all I saw of the Pittsburgh office.

HKS: But, as it turned out, you were just about out of a job anyway. Construction, of the magnitude you were involved with, more or less stopped when the mill got going?

RJG: I could have hung on. But, personally, I could not have taken it. With the Brazilians coming in with their pompous attitude and taking over the project, and with my temperament, I could not have taken it. Johan stayed to the bitter end. I could not have done it. I mean, I would have—I would have shot somebody or got shot myself.

HKS: I see.

RJG: To this day I remain extremely bitter about the way that the project was taken from Mr. Ludwig. He invested a billion dollars in the project, he put about seven thousand people directly to work, and forty thousand people indirectly, and was the major contributor to the economy of northern Brazil, and particularly the state of Pará. Then to kick him out like they did, and never invest another dollar of their own, just take it from him like they did—I remain bitter to this day. I haven't gone back to the project since I left, and that's because I made a promise to myself, which I may not keep forever. I said, [laughs] The only way I'll go back is at the head of a mercenary army, to take it back.

HKS: There'd be headlines there, too.

Typical housing for professional employees. Thompson Photo.

Quarters for unmarried technical staff. Welker photo.

Tauary tree in native forest. Welker photo.

Burning debris three months after logging native forest. Thompson photo.

Containerized pine seedlings ready to be planted on burned area. Welker photo.

Hand planting pine in burned area. Welker photo.

Pilings in place to hold pulpmill and power plant. NcNabb photo.

Roadbed construction for both rail and trucks.. McNabb photo.

Brazilian worker. Welker photo.

Eucalyptus nursery. McNabb photo.

Big stick loader. Welker photo.

Gmelina logs ar the Port of Munguba destined for Finland for pulping test. McNabb photo.

Power plant under tow. Welker photo.

Power plant and pulpmill in operation. Welker photo.

Collection of wood specimens. McNabb photo.

Daniel K. Ludwig school in Bandeira. When the project started, Bandeira was a small village of maybe eight houses where Brazil nut gatherers lived. McNabb photo.

Eradicating leaf cutter ants. Welker
Photo.

# JOHN C. WELKER

## Introduction

John C. Welker earned a bachelor of science degree in forest management
from Auburn University in 1970. While working on a master's degree in
forest science and economics at Yale University, he volunteered to spend a
summer at Daniel Ludwig's huge forest plantation along Brazil's Jari River,
as part of his overall education. He wound up getting paid—to him a bonus
on top of the incredible experience—and he returned to New Haven and
completed his degree in 1972. Shortly he was back in Brazil as manager of
forest management planning for *Jari Florestal e Agropecuaria*, as Ludwig's
enterprise was officially called.

Just think of the challenge—and opportunity—that the fresh-out-of-school
Welker faced. One-quarter of a million acres of native forest were in the
process of being replaced with plantations. The three species of trees
planted—gmelina, pine, and eucalyptus—either were not natural to the
equator or little was known about them. To that add the training of an
indigenous work force and the development of a town and overall
transportation infrastructure in a remote part of the world. All of the pieces

had to mesh to produce adequate fiber to feed a pulpmill that would be installed. The plans had to work!

John's responsibilities expanded in 1978 when he became manager of forest operations. Now he would coordinate forestry with logging in line with the needs of the pulpmill, sawmill, and the wood energy plant. In 1983 he left Jari and returned to graduate school, this time at North Carolina State University where he earned a Ph.D. in economics and forest management in 1987. Since then, he has been employed by Mead Coated Board in Phenix City, Alabama.

# Life At Jari

John C. Welker (JCW): After I graduated from Yale, I asked them if they were looking for anybody, and I applied for a job. They accepted my offer, and so that's when I started to work for Jari in 1972.

Harold K. Steen (HKS): Were you married?

JCW: No, I was not.

HKS: O.K. That made it easier in that sense. Was the financial package at Jari attractive as opposed to working elsewhere?

JCW: I didn't make any comparisons. My motivation was primarily to work and see the relationship between forestry and economic development, so this was just an ideal situation. You might have even looked at it, probably the way I perceived it subconsciously, as just an extension of my education. I wasn't married, I was young. My utility function, as an economist would say, was modest. The pecuniary motives weren't really there. I was more interested in the experience.

HKS: What were the benefits, other than feeling good about yourself?

JCW: The experience. Coming at it from the perspective of somebody interested in forestry and economic development, here was a situation where you were basically three hundred miles from the nearest metropolitan area, in the middle of the jungle. You were basically using forestry as a tool to foster economic development, starting at the ground floor from growing trees all the way up. The experience. That was what was in it for me.

HKS: Did you pay rent for the place you lived in, or was that part of the compensation?

JCW: That was part of the compensation. When you comment about what wage rates were, the fact that we didn't pay rent and also, when I went there, didn't pay for my food either, as I recall, or if I did pay, it was very nominal and not any real deduction from one's net pay. That was a good deal. Your saving rate could be very high, because there wasn't anyplace to spend money. All there was was a company grocery store. We ate in the company mess hall, and if all your meals were provided, what were you going to spend your money on? Nothing.

HKS: I realize it may have changed over the years as the infrastructure developed. Was there a liquor store? Were there bars? Were there those sorts of things? I've read about hospitals and schools.

JCW: I'm basically a teetotaler. It seems to me that there wasn't much in the first years. When people would come to visit the project, they would bring in duty-free type stuff. You could buy stuff in Belém, *cachaça* rum. That was all freely available, the Brazilian liquor, and spirits, and so forth.

HKS: I guess it was pretty impressive by the time that you left, in a lot of ways, with hospitals, and schools, and playgrounds, and whatever. But since you were single, you might not have paid as much attention to that as you would had you been married.

JCW: From the eleven years that I was there, you could certainly see differences. And I was married two years after I got there; I started paying attention to those things more. When we first got there, across the river there were just a couple of houses occupied by local folks. The economy had been where the people would go out and collect Brazil nuts, pelts, and that sort of thing, and trade them for staple commodities like sugar and things like that with the local land baron or whatever. Then Mr. Ludwig came in and established a landing strip near the river, which we eventually moved up to the plateau (*planalto*) area, put in some company houses and the company mess hall. That was basically it, with the supermarket. That was the only place you could buy anything, this one supermarket. The commerce on the other side of the river—even though Mr. Ludwig owned the land—was in the federal territory of Amapá. He stayed off that side of the river for a number of years. So the local shacks and houses and commercial activity there developed before we had a bigger supermarket in Monte Dourado, the town.

HKS: Was Monte Dourado already established? That was the place, obviously, where the town was going to be developed. By the time you got there, those decisions had all been made.

JCW: Yes. In '71 it had already been called Monte Dourado. In fact, they had already moved the airport. I guess the first airport, back in '67 or whenever, had been right where one of the streets is in the town of Monte Dourado today.

HKS: I have a AAA map for South America, and Monte Dourado is on the map. It became a real place.

JCW: Yes. That's right.

HKS: Bob Gilvary observed that it was in the wrong location. It should have been built where the pulpmill was, because you had to transport all the labor eighteen kilometers.

JCW: I guess what Bob is saying, it would have been better to have the labor closer to the mill, from that perspective. On the other hand, the advantage of having that town away from the mill is it's a more amenable place to live rather than being right near a factory facility. In fact, now the factory for doing the kaolin processing is near the pulpmill. They've got a small town or villa right near the pulpmill. I was down there about three months ago, for the first time in several years, and I wouldn't want to live in that little village. I would rather live up in Monte Dourado. From the standpoint of industrial facilities, yes, I agree with Bob. It was far away. But from the standpoint of living, it was good that it was located farther away from the pulpmill.

HKS: By the time you were married, two years after you got there, there was housing for married couples?

JCW: Yes. There was housing for married couples when I got there, too.

HKS: Describe your house.

JCW: I may still have the original plans. Bob Gilvary designed this, by the way. It was a T-shaped house, where you had bedrooms at the top of the T, two bedrooms. You had a bathroom in the center, you had a living room in the top part of the base of the T, and then at the bottom of the T, you had your kitchen and the washroom. It was made out of wood, probably Brazil nut wood, and the windows were not glass windows. They were louvered windows with screens, and then you could close the louvers to keep the dust or the rain out. Each of the louvers was about four inches wide. You would close them by hand, and so you had the experience that you were almost camping out, because when it rained, the humidity was just tremendous in the house. But then that was kind of the bad thing about it, the mildew problem. The good thing was that you had this nice sense of air throughout the house. We had an air conditioner in our bedroom, which was nice for the hottest days, but at night it was fairly comfortable without an air conditioner, because things cooled off quite a bit when the sun went down. We did have twelve hours of night, you know, because we were right on the equator, so there was plenty of cooling effect. When it got hot was, say, at noon time, and particularly in the dry season when you didn't have a rain, that kind of cooled things off. There was a tile floor. It was very adequate and comfortable. If you talked to my wife, you might get a different perspective. [both laugh]

HKS: The relations between engineering and forestry when you arrived. Who was running the show when you got there?

JCW: By the time I got there, foresters were running the show. I think the oldest plantations we had were gmelina plantations established up on the plateau in 1967. That was a kind of an engineering approach. They went up there with bulldozers. It was a very crude concept from a forestry perspective.

Mr. Ludwig found this tree, by one of his persons that went out and did these sorts of things in Africa. It was actually a tree native to India and Southeast Asia. This fellow that Mr. Ludwig had going around saw this tree growing in Africa that was used for pit props. And it grew very quickly. He got some seeds. Bring the seeds over to Brazil, clear off an area with tractors, plant them, presto! You got a big tree! I mean, that was the concept. We recognize in forestry that soil is a very important component of growing trees in that environment, and that was not recognized in the engineering approach. To answer your question, it was a forestry directed program by the time I got on board.

HKS: Clayton Posey had been there several years. He was hired as a geneticist.

JCW: Yes.

HKS: But he obviously had moved beyond that rather quickly.

JCW: Clayton, it seems to me, got there in '70 or certainly '71.

HKS: Were most of the technical people that worked at your level from the U.S.?

JCW: If you're referring to technical in terms of mensurational, growth and yield kind, yes. I would say so. Staff sort of things. If you're referring to foresters and forest management, they were typically Brazilians.

HKS: What help did the company give you to learn Portuguese?

JCW: They didn't really give me any help. I didn't ask for any help. I'd had four years of Spanish starting in the seventh grade and went through the tenth grade, but I'd forgotten a lot of it. But the good thing was, I remembered enough of it that I really knew the structure. The syntax in Portuguese and Spanish are very similar. There's a lot of confusion, though, and differences in some words. Well, I'd forgotten enough of my Spanish that it didn't get me confused when I learned Portuguese. But it really helped me in terms of grammar, in terms of learning syntax and conjugation. That was always one of the problems with people who had Spanish as their primary language, who'd come from other Latin American countries—because for many of them it was very difficult for them to make the transition. They would mix up Spanish and Portuguese words.

HKS: How fluent did you have to be to be effective, to do your job? Could you carry on conversations, or just give instructions?

JCW: It's like communication anywhere. The better you are at communicating, whether it be in the same language or whether it be in the other language, the better, more effective you're going to be. I always felt it was essential to learn Portuguese. I'm a pretty verbal person, and so I think it's important to be able to talk. I enjoy learning a new language and getting some of the nuances between one language and another, the way things are expressed.

HKS: Were there night classes?

JCW: I didn't take any. I know some people were offered classes, particularly some of the engineering people that might have been down there for a shorter period of time. Some other people had the Berlitz tapes, and so forth. The people that I hired subsequently, I would recommend to them to get a good grammar book and to immerse themselves in the language. For example, on my first boat ride, where I had a lot of fun, I was there with a lot of folks who were basically just laborers going out there to work. It was really fun because we had that bond. I carried a notebook around, and I would draw a picture of something and ask them, How do you say this? And they would say. And we just had fun learning it. Always my recommendation was, to the people that I hired, use the language as a way to establish relationships. Here's one of the things that the people, the local folks, know better than you know, and that will make them feel good, too. They know something you don't know.

HKS: Sure. Clayton made a point how important it was to learn the language, that he would fire people if they didn't learn the language after six months or something.

JCW: I probably learned the language faster than most Americans that came down there. They always kind of used me as an example, that even somebody as stupid as I can learn the language.

HKS: [laughs] I don't know how similar Brazilian Portuguese is to the kind they speak in Portugal.

JCW: There's a difference between northern Brazil and southern Brazil, too, quite a bit. But it's similar enough that you can understand one another. In Portugal it's a lot heavier accent. During the time of the Angolan civil war, there was some recruiting done in Angola and Mozambique, and we had some of those expatriate Portuguese come to the project. That's where I noted some of the differences.

## Daniel Ludwig

HKS: How many times did you talk to Ludwig himself?

JCW: I talked to Ludwig maybe twice the whole time I was on the project. I want to say it was toward the end of Ludwig's involvement with the project. They asked me to take him on a field trip and show him some things, and I did. I can't remember a lot from the field trip, you know. It's always the case where you're sitting there talking to somebody with a very high authority and you don't know quite what to say. It really is up to that person to say a lot to you. But I think we went out and looked at some gmelina plantations and so forth.

HKS: The descriptions of him in various kinds of media, *Fortune* magazine and whatever, describe him as a tall man. Yet there's a photograph of him with Ronald Reagan, and Reagan is a good six inches taller.

JCW: I don't remember him as particularly tall. As he got older, he stooped over a little bit like most people.

HKS: You don't have a sense of a dynamic personality or a tough minded person.

JCW: I do have a sense, I think, of a fairly tough minded personality. I'd been in some meetings where Ludwig was there, you know, and I was just an observer, a bystander.

HKS: O.K. I'm fascinated with Ludwig because all of the accounts, except maybe the one that you coauthored in the *Journal of Forestry*, are critical of Ludwig, of his vanities, he didn't like the press. You begin to feel that one of the reasons there are so many negative articles about Jari is that people wanted him to fail. It's incredible, when you read the twenty or thirty articles I have, that the guy was a nut. Jungle madness, on and on and on. You're the third person now I've talked to who actually knew him, and you don't describe him as a nut. His disdain towards academics, his short temper, and all.

JCW: I think he probably did have a temper. I think he probably got frustrated a lot. But I can't say I knew him enough personally to know what his character traits were. My perception of him is, as I say, he's a doer. He was a person who had amassed a large amount of wealth by going with some gut feelings after World War II, in terms of the oil tanker business, and so forth. Like most of us, he was a product of his own successes and his own failures in life. I've described his simplistic concepts of how to go about developing plantations in the tropics, which over time had to be informed with some forestry opinion and forestry expertise. But still, he had that basic idea that we need to move on and get things done quickly, because he wanted to see this stuff before he died. He wanted to see the fruits of his labors. If you think about it, you would say, well, if he really wanted to see the fruits of his labors, maybe he shouldn't have thought of forestry as where to do it. What was always told to me by Clayton Posey and others, was that Ludwig's basic vision behind why he wanted to do this in Brazil, in forestry, was he thought there was going to be a worldwide fiber shortage in the 1980s.

HKS: O.K.

JCW: His desire was to go in and create fast growing plantations and help solve that. Well, first of all, there wasn't a worldwide fiber shortage in the 1980s. But secondly, forestry, even if you've got short growing plantations, is still a long-term investment, because there's a learning curve you go through. I don't believe that Mr. Ludwig appreciated the type of learning curve you need to go through in an agricultural crop in a new environment. I think that's where the frustration came in, and maybe some people were very critical of him, and to some extent rightly so, in not recognizing that nature of a forestry investment. But on the other hand, where I respect the man is his ability as a man of action, to be willing to take his assets that he had earned and, with a hunch or whatever, to go forward with that hunch and put his money where his mouth is and not waffle. So I have to respect that to some extent. I think in the long term, a lot of those short-term things like return on capital and stuff will be forgotten. It's a bigger issue than just that in the scope of history.

HKS: You've already said you didn't remember very much about what happened. But when you took him on this field trip, did he seem to be learning, or was he dissatisfied by what he was seeing? The trees weren't growing fast enough?

JCW: When I first got down there, Mr. Ludwig would come fairly frequently to the project, let's say once every four months or so. As time went on, those trips became more infrequent, probably in part due to his back problems, probably in part due to some other business things, and probably in part due to some of his frustrations with the project, eventually leading to his pulling out. When I met with him, it was toward the latter part of this. He was down there for one of those infrequent visits, and I'm sure he had a lot of things on his mind. So what I remember was that he was pretty much preoccupied with other things. Even though they wanted me to take him out and show him the gmelina plantation.

## Typical Day

HKS: Describe a typical day. How did you know what needed doing? What kind of a plan was there?

JCW: I had several job responsibilities over the total time. My initial job responsibility was to set up a continuous plantation inventory system with the objective of not only knowing how much wood we had out there, and how fast the trees were growing, but to be able to project forward, create yield tables, project volume forward in time. As part of that, setting up the continuous plantation inventory system in the first years, I basically had three people working including myself. I had one person who was Marcos Braga, a forestry technician I think from Santa Marta. Basically a two-year agricultural school technician. By the way, Marcos later became a logging contractor, so he became an entrepreneur in his own right.

The other one was a fellow I think named Cristovam, who probably had a sixth grade education. That was my work crew. I set up the sampling design for this plot system, and we would go out and measure the trees. One of the real advantages was that I got a real feel for what the trees were doing in the ground, because I had visited every one of these plots. Now later on we had more and more plots, and my job changed. I didn't get out in the field as much, but I really got a sense of what was happening on the land base.

So my typical day would be—let's see. We started work at seven o'clock in the morning, and we finished up at five o'clock in the afternoon. We'd get in the pickup truck and go out and measure plots and trees. At the time we didn't have personal computers, and all that. In fact, one of the big deals was where to do the data processing. Later on we worked out an arrangement with Oklahoma State University, where we actually did our data processing up there. Gradually we did that in Brazil, as we got some other computing resources.

I'd start work at seven o'clock or seven-thirty and usually eat at six o'clock at the company mess hall. For lunch you'd generally eat out in the field. There were some camps for the laborers out there where you'd have black beans and rice. Then you'd come in and you'd eat in the mess hall at night. That was during my bachelor years, from '72 to '74.

HKS: You had a routine.

JCW: That's right. On Saturday we worked from seven-thirty to ten-thirty, although we typically worked longer than that. You've got an environment where there's nothing to do except work and sleep, and of course you did have social activities with the other families or whatever, but work was pretty much your life.

HKS: What did you do during your leisure time?

JCW: I read, and I liked to go hiking. I went fishing. I remember there was a fellow named Jairo. He was a mechanic, a Caterpillar mechanic. Brazilian. I went fishing with him several times.

HKS: Who did you report to?

JCW: I initially reported to Clayton, and then later on I reported to Charles Briscoe. Initially, there was just Forest Management. That was the period I reported to either Clayton or to Charles Briscoe. Then when we started gearing up for harvesting, we developed Forest Operations. Under Forest Operations you'd have Harvesting and you'd have Forest Management. I reported to Briscoe for awhile in that, but then eventually I reported to Johan Zweede, who was responsible for Forest Operations. That was basically the person I reported to until I left the project, Johan Zweede in Forest Operations.

I had at maximum two expatriates work for me at a time. We would look for an expatriate with respect to biometrics and statistics. One of the problems was in the Brazilian forestry schools, there wasn't a very good supply of people who were foresters who had good enough knowledge of statistics and biometrics, quantitative aspects of forestry.

HKS: O.K.

JCW: Initially those were the sort of people I would hire from the United States. Forest economists, statisticians, biometricians. As time went on, the Brazilian schools started graduating persons proficient in those fields. Going back several months ago and visiting again, I think there are certainly very qualified Brazilian foresters, just as qualified as they are in the U.S. right now. It's really nice to see the Brazilian educational system come up to world class standards in terms of producing foresters for the type of forestry that is practiced there.

HKS: It must be some sort of a market response. There were jobs for those kinds of students, and so they were able to...

JCW: Industrial forestry has really come a long way. In fact, I think Brazil in terms of industrial forestry, plantation forestry, is in many respects ahead of the U.S. right now.

HKS: Clayton talked about this, and I could tell he was a bit bitter by the general reaction of the outside world. He called them the propeller tie scientists from the U.S., the professors who'd come down for two weeks and then come back and write an article about it. There was very little publishing that came out of Jari at the time.

JCW: Yes.

HKS: Was it something that you observed too?

JCW: We had some frustration with people who would come down, and maybe it had to do with the persons when they visited, their motives. You always have this thing when the journalists come down, what you think their motives are. As you have more experience with journalists you get a little more wise to things. But whether it be a journalist or whether it be a scientist, forestry is, like most biological sciences, an applied field where it's very important to be close to the project at hand and to learn the problem. Just like Mr. Ludwig made the mistake of thinking that you could plant these trees and they would grow.

I think that scientists, armchair scientists, particularly ones new to their work, can make the mistake to think they can fly into a situation and in twenty-four hours, or a very short period of time, really understand what's going on. I was also frustrated sometimes when people would come with their own preconceptions and leave with those same preconceptions without listening and not being there a long enough period of time to really understand the problem. The problem is that they really weren't there long enough to understand the problem. O.K. Now I think some other frustration would be that the objective function or the motivations behind the people were different; it was just a difference of opinion. I mean, it's not that one person was right and the other person was wrong, but people just had a philosophical difference of opinion of the way things ought to be.

HKS: O.K.

JCW: We just had different opinions. As we reacted to what people would say in the press about the Jari project, it was just a difference of opinion. For example, is the Jari project a model for development in the Amazon? I can make a case and say, Yes, it is a model, because you're trying to, on a smaller piece of land, go to intensive management so you can preserve land elsewhere. Other people can make the case and say, No, it's not a good example, because we need to preserve the entire Amazon.

HKS: O.K.

JCW: But it's just a difference of opinion. So I think there was some frustration perhaps on all our parts when people didn't see things the way we particularly saw them.

HKS: But you didn't feel compelled to publish in the *Journal of Forestry* or other outlets.

JCW: When I was describing my typical day—we were so focused towards our work. Maybe it was Mr. Ludwig rubbing off on us [laughs], but we were so focused; we would be given objectives—to get so many acres planted, or here this pulpmill's going to start in 1980 or 1981. Those were the things we had to get done. We didn't see our role down there was to necessarily publish, at least initially, to publish our results. Later on, we became more outgoing in trying to publish some things.

The other part is the physical remoteness of our project. When I first went down there, there was only one plane a week, or maybe it was two times a week, and one boat a week. As time went on you'd have maybe one plane a day. That doesn't lead itself to publishing or contact with the outside world. I particularly know now in my present job, when you can pick up a phone and talk to somebody, or you can get on the Internet, that leads to a kind of interchange. So I guess there were two reasons we might not have published. One is because we had other goals in mind, and secondly because we were isolated, and it wasn't in our work objectives to publish necessarily.

## Applied Research

JCW: Off tape, you were commenting about how Bob Gilvary referred to some sort of micromanagement. He may not have used that term, but some of the sort of things that Mr. Ludwig did with respect to engineering. In forestry—maybe in the beginning it was a little bit of micromanagement in the sense of the idea of, we're going to select this tree species.

HKS: O.K.

JCW: I mean, that would be micromanagement from the standpoint of that's an important decision which shouldn't be taken lightly. You should do studies and all this. And so Mr. Ludwig micromanaged in that instance. Another kind of example was the decision to grow pine trees. Clayton Posey probably told you that Mr. Ludwig didn't want to grow pine trees.

HKS: That's right.

JCW: Clayton went ahead and planted some pines back in 1970 without Mr. Ludwig's knowledge, on an experimental basis, and then showed Mr. Ludwig this later. That was what brought us to eventually plant pine trees beginning in 1973. I think as time went on, as Mr. Ludwig probably recognized he was out of his depth and out of his field, he kind of left us alone with respect—us meaning forestry—alone with respect to micromanagement. Now he didn't leave us alone in terms of questions like, you're going to plant this many acres. Which again could be interpreted as micromanagement, because you probably need to put that through a little more sophisticated model, as to how many acres you should plant. But he didn't get down to the micromanagement about how you should do your nursery or anything like that in forestry, which apparently he did with respect to equipment decisions with the engineering side of things.

HKS: Ludwig would see the forecast, you're going to need so many cubic meters of wood, and you have so many cubic meters per acre. I don't know where he would have gotten ideas that he needed to clear more land, other than obviously you'd have to clear more land if you're going to have plantations.

JCW: I wasn't in on a lot of those conversations as to how many acres to plant exactly. Clayton was more involved with that than I was.

HKS: How did you know how to do things? Like, when you did your yield tables. My perception is that it wasn't a well known process how to do yield tables for very young plantations. The tree has to be a certain diameter before it has any volume. What models did you have?

JCW: The kind of theoretical models that I used were the models from the literature for pine plantations and for other things. You know, kind of the basic equation forms. But you're right. You have to change your perspective in terms of the information—what information you have to build those models with, or to use those models with. There's some real advantages in the tropics, in the Jari situation, because your information comes in quite a bit quicker. Whereas with, say, loblolly pine, on a twenty-year rotation, you probably have to wait till it gets to be at least twelve years of age to start really getting any kind of information about yield. With gmelina, which peaks out its mean annual increment at about six or seven years—at least in our conditions it did—at about three years of age you start to see some definite trends. That means you could put an experiment in, and if the experiments were well located, you could start getting some feel. In fact, I would say our problem in yield projection with gmelina was not the yield projection for a given site, although we had some initial problems with that, but really knowing how many acres we had of those sites. And knowing the right site classification. That was really our yield projection problem, because gmelina turned out to be a very site specific, site sensitive species for the soils that we had at Jari. And for that reason probably it was not a good choice for the long run, and subsequently has been basically abandoned as a species.

HKS: Gmelina in other areas, like Costa Rica and Africa, you didn't have that—the site issue wasn't—

JCW: Apparently they have better soils. In fact some of our initial volume information that Mr. Ludwig had about gmelina—and I guess I've sort of described earlier this kind of real naive approach. I should say, to give him credit, that Mr. Ludwig had some land in Costa Rica where he established some plantations, a plantation or experimental plots of gmelina. And so he had some knowledge of gmelina's growth. But it seems like that was established, say, in '66, so it was almost done at the same time the Jari project was set up. But to get to your point about sites—we had soils which were a lot more variable, I believe, and a lot less appropriate for gmelina as a species than Costa Rica, where apparently Stone Container Inc. is using gmelina right now. I haven't visited those plantations, but I think they've had some success there in growing gmelina.

HKS: The point I'm trying to get at here, the point of how isolated you are. You didn't have a major technical library.

JCW: We didn't have, for example, anything to go to for gmelina yield tables. We had to build our yield tables from scratch. You could find in the literature some things about gmelina, but it was more of a botanical nature or about silvics of the tree. It wasn't about growth and yield. Later on when we approached the problem with *Pinus caribaea hondurensis*, which was the other species that we developed, we had a little more to go on because that species was being grown in Fiji and some other equatorial regions. So we could get a little more out of the literature from that. But with gmelina we were really starting from scratch.

HKS: What access did you have to literature? These days you can go online, onto the Internet and you can dial things up.

JCW: We had a weekly pouch from the States, but we didn't have anybody really in the States that we could say, do a literature search in the library. So typically what I and some others would do when we were in the States on vacation would be to go to a library and spend some time. I can't remember when it started, I want to say in 1972-73, we established this relationship at Oklahoma State using their computer center. A fellow named J. L. Albert worked at Oklahoma State, and he did the processing of our growth and yield, our plot data. When I was up there, I would look in the library and I would do some interlibrary loan and that sort of thing, do some literature searches. Another source was Oxford, I mean in the sense that *Forestry Abstracts* being more of an international type thing, and because of the British influence in Africa, that would be a place you could find a little bit of stuff on gmelina, for example.

HKS: I've been told that the largest tropical forestry library in the western hemisphere is the Forest Service library in Puerto Rico. Did you make any use of that library?

JCW: No, we didn't.

HKS: So you would go to a forestry school in the States that was convenient.

JCW: I would tend to look in *Forestry Abstracts*, which of course covered whatever had been published. You'd look for a species. This all seems so antiquated compared to today's search techniques.

HKS: That's something you actually received by mail. You had it right there in Jari.

JCW: We developed our own little library there at Jari, which subsequently burned up in our office fire. Charles Briscoe had worked in Puerto Rico, for example, with the U.S. Forest Service, so he had a tropical forestry background. One of the things he was instrumental in doing was to set up the library there at Jari. So we accumulated books over time about various aspects of tropical forestry.

HKS: Clayton said that he had more people in research than he had in line work, at one time, in the forestry side. He didn't define what research was, but apparently it was a major aspect of the effort, because there wasn't knowledge.

JCW: That's right.

HKS: You had some empirical observations that you could draw upon...

JCW: Yes.

HKS: How about consultants? Did you bring in soil specialists or botanists? I'm not sure what kind of specialists you'd need to do the job.

## The Problem with Gmelina

JCW: We had specialists come in from time to time. When Mr. Ludwig would apply for loans, you would have to have consultants come in for the agency loaning the money to do some things. So you'd get consultants that way. We had a fellow named Ian Bailey, who was a subcontractor with Reid Collins. He was from England, came in as a soil person. We had Harold Burkhart and Zeb White who came in one time and looked at growth and yield. Frank Bennett, too. There were some kind of negotiations with Weyerhaeuser. We had two fellows, Alex Goedhart, who was with Weyerhaeuser out of Chehalis, Washington, and the other fellow that I can't remember. They lived there for about six months. I wouldn't describe them so much as consultants as much as, they were there to try to learn what was going on. They'd also do some consulting and give some advice, too. Charles Hodges came in. He was a pathologist in the Forest Service. When we were looking at *ceratocystis*, which is a disease for gmelina—

HKS: The chancre—

JCW: The chancre disease.

HKS: I wanted to ask you about that, because there's disagreement whether that chancre was really serious, and the paper that you wrote says that it was. But Clayton said that when gmelina was under stress, it had a problem, like any other species, under stress. The chancre was no more of a problem than it would be for any other species. I'd asked him the question because Fearnside and Rankin wrote that series of articles. As a matter of fact, you cited them in the articles you wrote.

JCW: Yes.

HKS: I assume that you find them to be respectable scholarship.

JCW: No, I don't. I don't know whether they're respectable scholarship—with respect to the Jari project, I have differences of opinion, of the way we see the world. That's just the way people are. We see things differently. I also have differences of opinion scientifically with them. In other words, the interpretation they give to some facts about Jari, some articles I've read, I don't agree with. I don't think they're correct factually. Judy Rankin, she came and visited. She asked me some questions, and we talked. A couple of years later, I saw this article, and it was as if we had not been with the same person interviewing.

HKS: I see.

JCW: I thought they took more of a journalistic approach to things. When I say a journalistic approach—getting back to one of these questions you'd asked me about what Clayton thought—I think they came to the Jari project with some preconceived notions. They had a hypothesis and they were looking for facts to support that hypothesis. As a scientist, it's all right to come in with a working hypothesis, but you have to take those facts which will support plus those facts that don't support that hypothesis.

HKS: Absolutely.

JCW: I believe that there's a tendency in some of the stuff I've seen for them to disregard the facts that don't support their hypothesis and only accept those that support their hypothesis. If my perception's correct, I don't agree with them. Oh, the chancre disease.

HKS: The chancre. Rankin said it was a serious problem.

JCW: I would agree with him on that. I can't remember the exact year that the chancre disease surfaced as a real management issue for us. This is a disease that's on cacao. It's a native, it's an endemic disease with cacao. It has a pretty large range, or the potential to have a large range. The scientific name at the time was *Ceratocystis fimbriata*. But to counter what Clayton is saying about stress, we had some of our biggest problems with the chancre disease on the best soils, in other words, on the terra roxa soils. My working hypothesis of where it got its start on our trees was when we had a pruning program. Charles Hodges would be the person to talk to about the scientific aspects of this disease, but it comes in on pruning scars. Now subsequently where it would come in was in the coppice management of gmelina. We weren't really wed to coppice management, because you can't capture genetic gains with coppice management. But when you cut the gmelina it does sprout back profusely. In fact, that was one of the problems—how do you kill this stuff so you can put in your second generation? Or, if you want to convert to another species, how do you get rid of the gmelina to plant pine trees? Because it's hard to kill with the herbicides we had.

But anyway, with that coppice management there, you have areas for the spores to get into. From our growth and yield plots, our continuous plantation inventory plots, we started recording for every tree whether it had *ceratocystis* or not. I worked up a matrix, a transition matrix. Like if a tree has the evidence of it, what's the probability it will die in the next year.

HKS: Sure.

JCW: One of the things that became very alarming and one of the reasons we started really shifting more aggressively looking at eucalyptus as a species in the latter years was that those transition matrices were saying that our mean annual increments would start peaking earlier and at a lower rate. I would say it had a really significant effect, that and soils. The other problem with gmelina of course, as I said, is its yield goes down tremendously with poorer soils. I think gmelina probably has a place at Jari, if the management wanted to continue it with the genetic testing. But the question you have industrially is, do you want to bother with it?

HKS: Sure.

JCW: In other words, this is a cost. You're going to have to improve it.

HKS: Chancre may not be a good example, but I'll use it for lack of specific knowledge.

JCW: Sure, that's fine.

HKS: You go out one day and you see something. You have a general forestry background, you're certainly not trained in pathology. And you see this thing. You don't know if it's serious or not. You probably saw lots of things, most of which turned out not to be serious. Where was the expertise that you would draw upon? Would you bring in a pathologist?

JCW: Charles Hodges was brought in. We had had Charles Hodges and Charles Briscoe, who had known each other in the Forest Service.

## Eucalyptus

JCW: Let me talk a little bit about eucalyptus. We had research plots with eucalyptus going back to 1970. *Saligna* and *grandis* and so forth. In retrospect some people might ask, Why didn't you plant eucalyptus? Everybody else in Brazil was doing it. They're doing it right now at the Jari project. We started planting it. One of the things that scared us with those eucalyptus plots was that they had a chancre disease. Not *ceratocystis*, but another one.

HKS: O.K.

JCW: By the way, Brazil has been able to conquer that disease by looking for the right genotypes, the right species, that are resistant to that. Probably in retrospect I would say we should have pursued that more vigorously in our genetic programs; eucalyptus is an important species in Brazil, and we're in the Amazon. Everybody else is really not in the Amazon. They're in subtropical regions. Now I think *Eucalyptus urograndis* and some others are successful, and those genotypes are being developed at Jari. They're getting very good yields there, which are, I think, going to be competitive with the rest of Brazil. Now we did adopt a strategy back in 1979, where we started doing eucalyptus species trials, looking for other species. On that basis we looked at *Eucalyptus deglupta*, which is resistant to that chancre. It is one of the true tropical eucalypts, not the subtropical but true tropical. We actually started planting *Eucalyptus deglupta* on a wide scale as a way to sort of meet some fiber shortfall problems because of the gmelina problem.

HKS: The literature on eucalyptus was probably pretty hefty.

JCW: Yes. That's right. If you look at the bands on the earth of where successful plantation projects are, they're not on the equator. Aracruz, for example, is often cited as a very successful project in Brazil. That's really not in our zone. That's a totally different zone, different rainfall patterns and everything. There are not a lot of projects right on the equator, and eucalyptus is really a subtropical species, on the whole, more than it's a tropical species. Now as I say, Jari is having success with some of the subtropical eucalyptus species. But that's largely due to genetic testing and finding those genotypes within that species that can do well.

While I'm on this digression about genotypes, with respect to pine, the current project, right now, they're weaning themselves away from pine. I think there's tremendous potential for developing pine as a tropical species, but only with a lot of genetic testing, because if you look at the plantations, there's a lot of variability that can be captured. Now whether Jari or any other company wants to capture that, if that's a good business decision, that's another point. If I was to look at two keys for setting up industrial plantations, it would be to pay attention to soils and pay attention to genotype, and manipulate those two resources to your greatest advantage.

HKS: Well, Clayton, the boss man, was a geneticist.

JCW: Yes.

HKS: And he must have been more sensitive to this than the typical manager would have been.

JCW: Yes.

HKS: But even then, it still wasn't enough. Looking back you see there should have been more done.

JCW: The thing that I keep going back to, it's getting back to the character of Mr. Ludwig and his time frame versus the time frame of forestry research. Even though forestry research can go a lot quicker in the tropics, it couldn't go as fast as Mr. Ludwig wanted to go. At Jari it couldn't go as fast as it could be in other parts of Brazil or Australia or wherever, because we didn't have the literature on gmelina like there was on eucalyptus in the subtropics. So we had to go faster and try to do some things right from the start. We had kind of two strikes against us, in terms of a learning curve. One is, we weren't given a lot of time to learn. Secondly, we had not as steep a learning curve because we had to do a lot of it ourselves, rather than get it from the literature and from what previous researchers had done.

HKS: Has enough been published on the gmelina experience that, should someone else want to try it, they could build on?

JCW: I don't know if enough has been published, because I haven't been in that literature. But there certainly is enough experience now, if you take the Jari experience, and there was some stuff published on that. Maybe not a lot. We did articles in the *Journal of Forestry*, but that really wasn't about gmelina itself. In terms of experience with gmelina, I think there's enough experience out there that's being carried on, whether based on Jari or based on what they're doing in Costa Rica.

HKS: At the time this was happening, gmelina was seen as sort of the wonder tree. You could do everything but eat it.

JCW: Yes.

## Lessons from Jari

HKS: Can you generalize from the Jari experience? If another Ludwig comes along and wants to do something, would there be a cookbook that could apply in Indonesia or somewhere else?

JCW: I don't think it would be a cookbook. I think that there's some things that someone can learn. A lot of aspects of Jari were so unique, kind of one at a time type things, that I don't know what you could get from that, for benchmarking for future projects.

HKS: A lot of the critics of Jari, and there's a lot of those comparing it to Henry Ford, mainly because there were billionaires behind both.

JCW: That's right.

HKS: Well, Ford couldn't buy nature, and Ludwig's not going to buy nature either. Nature's going to win. And look what happened to Ford.

JCW: Sure.

HKS: According to Clayton, there was a knowledge of what Ford did. He said it really wasn't comparable, the problem was that the genetic base of his rubber plantation was far too narrow, and it would have been wiped out anyway, or some such thing.

JCW: I don't know enough about that. I do remember hearing the two compared in the popular literature, newspaper articles. Will this be another Fordlandia, that kind of a spin on things. But I don't really know enough about what really happened at Fordlandia, whether it was economic or biological or a combination of the two.

HKS: There was a disease that wiped out the rubber trees.

JCW: O.K.

HKS: It grows O.K. naturally, but not in plantation.

JCW: With respect to forestry development, I guess the key learnings I would have is, you need time. You need a good research program. We had a lot of people devoted toward research. But I think you need a little more time than we had. If Jari had been a stockholding company, you would have taken a lot longer time to make some of your strategic decisions. I've heard people counter and say, Well, if it had been a stockholding company, Jari never would have happened, because the risk would have been too high for any stockholding company. You'd have to give Ludwig credit for being far-reaching, because even though there wasn't a fiber crisis in the 1980s, there was a need for more fiber. I was at a recent RISI conference, for example, and what they're pointing out, there's plenty of capacity worldwide from a manufacturing standpoint in the world, in the wood products industry. But the limiting factor is going to be wood, in terms of how fast that new capacity can keep growing. So in some sense, there's not a crisis, but there is a need to grow more fiber right now.

To repeat a bit, I think you do need more time, if you want to reduce the number of mistakes. And I think the role research has to play is not only to increase productivity, kind of its traditional role, but also to increase your tool kit or your arsenal. If you're going to go into intensive management of plantations, which is what basically Jari and most of the Brazilian model is, you really need to take an agricultural paradigm, where your research is always looking at your genetic base for your crops, whether it be between species or within a species, so you're not vulnerable to disease and other things. I think that's one of the key learnings, because the research can really help.

The other thing is to look at soils, that soils are your basis for growing these trees, and in growing different species. You need to make your species compatible with the soils you have, because it's very difficult and costly to change soils. Those are the key learnings, for me, forestry production-wise.

I think the other key learning, with respect to building an infrastructure, is one of the traps that really kind of sank the project from Mr. Ludwig's perspective. When he went into that project, he had the support of the government at the time, which was basically the generals running that country, or a cadre of generals. Over time as they went into the phase of *abertura*, or opening of the democratic process. I believe Jari was used, in kind of a populist type model of democracy, as a whipping boy for the local and national politicians. Here this foreign national is in there, you know, with our birthright. All the infrastructure was built by Mr. Ludwig, and the roads and everything else. He never could get what I consider the government to put in their fair share for supporting that infrastructure.

HKS: Interesting.

JCW: Now the new company, being a Brazilian company, I believe has gotten a little farther ahead in that respect. My feeling during this past visit is that even they are having problems getting the government to support the infrastructure burden. They pay taxes, but much of the taxes they pay goes to the state government four hundred miles away. It's not being plowed back into the infrastructure there at Jari, so that that company continues to have to pay for a lot of that infrastructure. Any private enterprise that goes into a project like that really needs to make sure what the government is going to pay for and what they're going to pay for, because you're going into a pioneer area. The government, being like most governments, wants to try to get the private economy to pay for as much as possible.

## Computer Based Forest Inventory

HKS: We've already talked about how do you know things, and the library research, and consultants, and so forth. Let's talk about specifics. What was a computer based forest inventory?

JCW: We would measure the trees in the field. Using a pencil, we would write it on some forms that had already been preprinted from the past measurement of those plots, by tree, the diameter and the height and those sorts of traditional inventory measurements. Product class, whether it had a disease or not, whether it had died or not. We would write on these forms. Those forms would be sent to the United States during that period, be keypunched, using the old keypunch cards, run on the mainframe computer, print out summary statistics of the plot level, by soil type. So we would know how the inventory was changing over time. But one of the main uses of that summary data at the plot level was to develop growth and yield predictions, because we had the same plot being remeasured over time. J. L. Albert in Oklahoma and myself, later Joao Borges, we were the ones basically that developed the growth and yield equations from those plots, plots using experimental data that we had on other places.

HKS: Each tree was numbered with a tag on it.

JCW: Yes. Each tree had a tag. There was a plot center, it was a circular plot, twentieth hectare in size.

HKS: There was a statistically valid model and all of that?

JCW: Yes. They were put in at random. One of the real advantages of using that information for our growth and yield was, as opposed to say experiments in the field, that those plots were treated just like the rest of the plantation. So you weren't getting something that had been treated in an extraordinary manner silviculturally. It was an operational based inventory.

HKS: So until you had that, everything was sheer guesswork.

JCW: Yes, but we started that inventory system back in 1972. Putting plots in all plantations that existed from 1969 onwards.

HKS: Yes. I'm not sure when Ludwig made the decision to have a pulpmill.

JCW: By that time we had already had the inventory system in place. I want to say that the decision to build a pulpmill was probably made about in 1977, '78, right around then. Because one of the reasons he floated that mill across was so he could get done quickly. They probably cut the project time by two years by doing what they did rather than trying to build it all on site. I would say about 1977, maybe '76, was when things started moving towards when that pulpmill would start.

HKS: I'm assuming that a pulpmill has to have a certain production capacity in order to be viable. So many tons per day or whatever it takes. You've got to start with that. It has to be at least that much.

JCW: Right.

HKS: Then you have to say, Well, can I get enough stuff off the land?

JCW: That's right.

HKS: I've read that there was still pulp imported into Jari, from southern Brazil, because there's not enough local production. That was in the '80s, '70s. I don't know if that's still the case.

JCW: Yes.

HKS: Was that a miscalculation? Or just good business practice?

JCW: No, it wasn't good business.

HKS: Was this a surprise when you didn't have enough to feed the mills?

JCW: How would you describe this? It was a combination of factors, one factor being the disease problem with gmelina. So that reduces yield. Then you start looking for something to fill in some gaps. So you look for eucalyptus. We chose *Eucalyptus deglupta*. Now *Eucalyptus deglupta* probably was not the best choice. It may have been the best choice given the information we had about the genotypes that were available to us. Then you get the eucalyptus to fill in the gap. Another way you could have responded to that information about the yields on gmelina and then the *Eucalyptus deglupta*, would be to say, Well, we're going to plant extra acres to perhaps make up for that possible shortfall, though of course it's still uncertain even if you've got a shortfall. As a risk management type of thing, let's plant some extra acres. Then you're getting into the period where there's the transition to the new ownership, where Mr. Ludwig no longer owns the project essentially, or he knows he's not going to own the project. So then you say, does it make sense for us to plant risk management acres, when we're not even going to own the asset?

HKS: O.K.

JCW: And we probably don't need to plant in order to sell the asset, and particularly an asset that we're selling and we're not getting top dollar for, anyway. So I guess you could put those three things into play, the *Eucalyptus deglupta* not doing maybe as well as the current eucalyptus does, the gmelina giving you a problem, and then not planting enough acres to say when you're reaching into a kind of possible problem of a volume flow. Then you get this new company on board—and there's a bit of a speculation in here, because I wasn't there after 1982 to know what goes on. But I have a feeling with the new company, they got into some cash flow things, they got into some of their own assumptions. I know, for example, there were some statements made, Well, we're really going to grow gmelina now because we're going to start adopting very intensive management, like in southern Brazil.

That kind of a contrast. They got caught in their own paradigm, and cash flow problems. So as I understand it, the project continues to struggle with fiber flow in part because of cash flow problems, and not planting enough for risk management. We're getting outside my boundaries of when I was there, but they were buying fiber from AMCEL, which was not too far away, which was partly owned by one of the other companies that owned the CAEMI company. That's not a bad decision, because here you've got a pulpmill that can barge wood over to you. It's a win-win situation, the way I looked at it at the time, because the closest source for that pine fiber is that pulpmill. I mean, it's the closest mill around. I don't think it's necessarily a good situation to be bringing wood from southern Brazil or Bahia and so forth, because that's a pretty long stretch, without a lot of backhauls and all the other things you need in shipping. But that would be my scenario of how the fiber flow problem unfolded.

HKS: Did you forecast that there would be a shortfall?

JCW: No, we didn't forecast there would be a shortfall.

HKS: I'm trying to put myself in your shoes. You're out there working every day. You know they're building this pulpmill that in two years is going to come online.

JCW: Yes.

HKS: It's no longer theoretical. I mean, there's a certain pressure on somehow.

JCW: Right.

HKS: If that's the right word.

JCW: Right.

HKS: This is not an academic exercise in forecasting.

JCW: That's exactly right.

HKS: It must have been kind of interesting from where you were with this.

JCW: Since you're in a situation where you're pretty close to a hundred percent, you have to have a hundred percent self sufficiency, which is a totally different aspect of the problem than we encounter here in the States. Here most mills are thirty percent self-sufficient. Some companies don't have any self-sufficiency. We were using some of the native species we identified that we could use for pulping. When I was there we didn't do it, but we could get wood from AMCEL, you know, a barge trip away. So there were some outlets, but not like in southern Brazil, to some extent, or in the U.S. for certain.

If you look at the yield projection aspect, I think it had to do with gmelina not performing, either because of the disease problem or because of the soil problem, to what we thought it could do. If you think of yield projection in the global sense, it has two aspects. There's how much, how well you can do projecting yields on a given acre of land of a given type. The other aspect is to know how many acres you have of that type. We didn't get into soil mapping. We used for our soil mapping a geologic delineation of the soils. Johan Zweede's area of interest and expertise was the geology of the land base at the project. We used geologic zones to identify our soils; we debated whether to go into an expensive soil mapping project, but we didn't start getting into that actually until after about 1982.

I guess the other aspect, in retrospect, is to think about how those plantations were developed and the technology. We were centered there in Monte Dourado, and you did have the ability to overfly an area. But aside from that, you had to go in with crews, camping out. Typically you would go out and you would do an inventory of the area, running lines in the native forest. Then you would make the decision, because the driving factor was how many acres to plant. So you'd go in there and you would clear the land that you had gone in by foot to get to. You'd build the roads after you had inventoried, and then you'd go in and clear it, you'd plant the trees, and you'd gradually just work yourself out from the town, being the epicenter. As you did that, you tried to get some information about soils. But we didn't do any soil mapping at that point in time. In kind of a more perfect world, with more time, what one would have done, is one would have established experimental plots over the whole land base, you know, building roads way out.

But we didn't build our roads far. I guess there are roads out there now that are maybe a hundred and ten kilometers away from the city. But we would just build our roads gradually and put experiments in as we cleared plantations. In an ideal world the roads would have already been there. You would have put the experimental plots out on the peripheries, or all over on all major soil types, and see how it responded, and done your soil mapping, figure out how many acres you had of all those soil types. Then you would have at least been a lot better on your yield projection. [laughs] But that's not the way it went.

HKS: Then you started there simply because it was a convenient place physically to build a town.

JCW: With a map I could show you where the soils are. I think in my article that I published I really talk a lot about this whole issue of the soils and which soils are better for growing gmelina and so forth. If you notice in that Puerto Rican article—the town is located on the worst soils for gmelina. Our best soils were over the mountain and up near the waterfalls on the project. That's probably a sixty kilometer haul, which by U.S. standards is nothing. We had to build all our own roads, you know, that's a pretty far reach up there.

HKS: I try to put myself in Ludwig's shoes. Various reports—he doesn't think a whole lot of academics, theorists, I mean. He's a practical man.

JCW: That's so. I agree.

HKS: He's going to make a big time decision based upon growth projections from three young, inexperienced people. He must have had a lot of confidence that you guys were doing something right.

JCW: He either had a lot of confidence or, I guess one could put another perspective on it, saying he was probably naive to begin with about forestry.

HKS: Maybe he didn't have any choices. What is he going to do?

JCW: That's right. Plus there were some people that had some experience. Take a Charles Briscoe, for example. He had had experience in Puerto Rico, he had experience in Mississippi. Now in terms of lack of experience at Jari. We all didn't have any. You know, we all were fairly inexperienced. We were inexperienced in that particular setting. He was willing to take a lot of gambles with a very limited amount of information.

HKS: How old was Briscoe at that time? I mean, was he considered a senior scientist?

JCW: I would say he would have been considered a senior scientist. I'd say he was in his fifties.

HKS: O.K. My perception was almost everyone except for the consultants were very young, in their twenties, early thirties. You, and Clayton, and Gilvary, and so forth.

JCW: Yes.

HKS: The senior people were the consultants that came in.

JCW: Yes.

HKS: The Zeb Whites, the Tommy Thompsons, and those guys.

JCW: I'd say twenties, thirties, and forties. I mean, there were some fortyish people in there, when I first got there. Again we're talking about a time stream in '72 through—

HKS: True. People aged.

JCW: Yes, it's hard to say.

HKS: I assumed that it was a hardship outpost, and people established in life were less attracted to it. You have kids in college here in the States, it's different than if you're starting out in life. You don't have a mortgage to pay in the States and all the rest of that, to take that job. That's sort of a selection process for those who actually went there.

JCW: In my experience in hiring people, not only in Jari, but in my current job, we talk a lot about motivational fit; how well a person would fit with that job, whether it be location or whether it be the job description. I would say that certainly someone young and single, you wouldn't have to worry as much about motivational fit as you would with somebody who's married with kids, because then you have to start asking yourself the question, Well, it's not only this person I'm hiring, I'm hiring their family.

HKS: That's right.

JCW: So you had to be a lot more careful in your motivational fit part of it, in looking at those people. I never was involved at that level of deciding, making decisions about some of the people who came to the project, but sometimes, I often wondered, at least in retrospect, how good a search process and how good a recruiting thing was done looking at motivational fit.

HKS: Recruiting would be kind of difficult.

JCW: It is.

HKS: The job market's back in the States and you're a long ways away.

JCW: That's right. It's an art in itself, even in the U.S. I see mistakes made all the time; people that are not doing a good enough job in the interviewing process.

## Management Information System

HKS: Talk about the management information system.

JCW: I guess there were a couple of things there. Early on we assigned each one of these areas that was cleared at a given point in time, what we called an absolute plantation number, which never changed over the life of the project. In fact, they're still using those absolute plantation numbers today. For example, you'd have each plantation being assigned a number, let's say number forty-seven. That was one up in the Pecanari area. O.K. Within that plantation you would have a fairly rectangular road system, depending upon topography. Each one of those would be assigned a number, a compartment number, one, two, three. O.K. We put that information in the computer, but because of our remote location, we also had files. When we did any kind of silvicultural activities, the area foresters would send in weekly to my group—what was done, where it was done, so we had a record. You can go over that rotation and say when it was planted, what the spacing was, when it was cleaned, how many times it was cleaned—meaning weed control—if it had been pruned or not. That was our management information system, to allow us to know silviculturally what was done.

At about half the rotation, we would do a stand level inventory, or compartment level inventory, to get basal area, height, which we would then use to project yields for that compartment. Those projections were what we were using to determine, when we did our harvest simulation, how much volume would be there to cut to meet our target volumes. We basically were using area control. As an example, let's say you were on a six-year rotation. You had sixty thousand hectares. When things got regulated, you would do ten thousand hectares a year, and that would lead to a certain amount of volume, and you'd get to the sustained yield. Now obviously, you never reached that theoretical state, but that's the way you sort of go about doing your volume forecast and harvest schedules. Today, that's now being put on a geographical information system (GIS) at Jari, because of what we can do now on microcomputers. The framework and the design of that system, even without the computers, was essentially the same thing you do today on a GIS system, where you have basically plantations divided up into compartments or stands, which you track over time with respect to yield and with respect to inputs.

In fact, I probably patterned it more off of what was being done in Australia and what you see in the British Commonwealth type literature with plantation forestry more than I did in the United States, because the United States really, until recently, and recently means the last twenty years, hadn't started tracking plantation yields like that. Now as we get into more intensive management, we're doing that, as we get GIS systems and so forth. But we had kind of a compartment registration system, as you would find in New Zealand or Australia. That's the way we did our management information system.

HKS: I was just thinking. When I went to forestry school, a plantation was an exotic concept. Probably something we might not even see in our lifetime. We were still old growth management.

JCW: Right. Right.

HKS: Here you were dealing with a six-year rotation.

JCW: Right.

## Tactical and Strategic Forest Management

HKS: Let's go on to the next topic. Tactical and strategic forest management. Harvesting plans. How do you coordinate this? Does Gilvary ask to buy the chainsaws and stuff, logging equipment, to keep up with the harvest?

JCW: No. No, Bob Gilvary—he was responsible for engineering, not forest engineering.

HKS: O.K.

JCW: He was the guy responsible for building the roads, building the town, building the water treatment plant, building the pulpmill, or helping with building, that sort of thing, civil engineering. Building the houses, the school, the hospital, all that sort of stuff. Going back to the way things evolved, we started with just forest management, growing trees. Then you got into the facilities planning type issues. At some point, Clayton Posey moved back to the States, and they formed a forest products group for Mr. Ludwig out of Stamford, Connecticut. That's where all of the planning and the decisions about what kind of pulpmill to build and all that sort of thing was done. At the same time, down at the project, we were gearing up, saying, What kind of harvesting system do we want? And hiring people to look at that issue, to get ready, because here from scratch, in the middle of nowhere, we were going to have to set up a harvesting system for the planted forest. Up until this time, we had done a little bit of native harvesting, but not much. There was a small sawmill on the project before the pulpmill started, for local consumption.

The first job, from my perspective, before we even got there, was to sort of do these volume flow projections, as we talked about, and some of the pitfalls and difficulties of that, and coordinate that. That would be strategic harvest scheduling, coordinate that with what the people in Stamford were asking, and saying, here's what we think the yields could be, but you're going to have to cut trees at a young age. Those sort of issues. Then as you come closer to actually cutting the trees, when we got online with the mill running, what I would call tactical planning is to make your one year harvest plan. In other words, what the question then becomes is, What stands am I going to harvest in 1981?

I worked with the people in procurement to give them some information on what the age of the stands were, and from stand level inventory what volume to expect off of each one of them, what the tree size was on them, because that determines harvesting cost, and help coordinate with the management part and the procurement part to know how much would be cut and how much needed to be planted after the cutting. The people in Harvesting, a fellow named Mac Davis, who used to work for IP, was the harvesting manager. His main staff support was a fellow named John Sessions, who's out at Oregon State now. When Mac left, John became in charge of Harvesting. On the Forest Management side—I think by that time Charles Briscoe had left, and I think Robin Collins, who'd worked for Westvaco in southern Brazil, was on the Forest Management side. Then you had Johan Zweede in Forest Operations, who was the boss of both those. Then my job in the tactical, strategic, and operations planning was as a staff support for Johan and for Mac Davis and for Robin Collins, to kind of help them coordinate their plans.

HKS: So Mac Davis would determine if he had enough workers, enough axes and saws, to actually harvest the amount of materials.

JCW: Yes, his job was to set up the harvesting system and the contractors.

HKS: I suppose Gilvary would make sure the roads were in place.

JCW: By this time the road system was pretty much in place, because we had to have the road system to plant the plantations. A major project that Bob worked on was the railroad, because that was an addition that was put in to handle the volume of wood coming into that mill. Ludwig at the time did not have plantations on the other side of the river, because that was the federal territory and he didn't want to develop over there. That changed subsequently. So the mill was located on a peninsula, and so all the wood had to come from the west. There was this sort of narrow funnel this had to come in, and one of the reasons behind the railroad was to try to not have so many trucks coming down the road. It was to help out a logistical nightmare.

HKS: An anecdote Clayton gave—early in the game, I guess when engineers were still running the show, he said they were clearing so many acres per year that the guy in charge of the nursery didn't produce enough seedlings to plant those acres. Lack of coordination. Obvious things, apparently obvious things now.

JCW: That's right. That was before my time. I'm speculating, but here you're talking about learning how to grow seedlings in a nursery, too.

HKS: Sure.

JCW: So it might have been that the nursery thought they were growing enough seedlings for "x" number of hectares, but then they had a failure, so they wound up with not enough seedlings. Who knows? But some of it could have been coordination, some of it could have been a downfall in nursery production.

HKS: Sure.

JCW: But by the time I was there, that problem was pretty much ironed out. I can't think of any time where we lacked seedlings to grow.

HKS: Was labor turnover a problem?

JCW: The labor in the early years for planting was from contractors, or from contract labor from part of northeast Brazil, where they really have a labor surplus and where people were looking for work. They would come up for a year at a time, or six months at a time, to work. That system inherently has a lot of turnover. But that's labor intensive, you know, planting crews. Now in terms of turnover, I'd say that where it was more of a problem, probably, was the turnover at the higher echelons, like the directors of the project changing.

HKS: Every six months or something.

JCW: Things like that. That was probably a bigger labor turnover problem, in terms of direction, than at the other end.

HKS: Do you want to comment on that, because I think there were thirty-four or thirty-three directors before Clayton. Some large number like that.

JCW: I don't know how many there were.

HKS: How did that affect you where you worked? I mean, the boss was maybe two or three levels above you.

JCW: It pretty much didn't affect me. I thought it was all just kind of interesting. First of all, one has to ask the question, Why would that be?

HKS: Sure.

JCW: I can think of three different reasons that it would happen. One case, and it gets back to this recruiting issue to a very large extent. This would be a person Mr. Ludwig would hire, obviously at this level. Mr. Ludwig's vision would be different than what the person's vision was, and there would be a clash of personalities, hence we know who wins that one. That would be one possible reason. Another one would be a lack of motivational compatibility. I think there were a couple directors that, they had some personal problems, that they were let go for that reason.

HKS: Clayton lived there. Did they live there, or did they live in Belém or someplace else and fly in once a week, or was that an issue?

JCW: For my time there, they all lived on the project.

HKS: O.K.

JCW: I think another problem is communications. Even though these people might have been able to work out a good relationship with Mr. Ludwig, there was the physical isolation. We didn't have phones there for most of the time I was there. Just to give you an anecdote, when I went to Belém in 1971 in order to make a phone call—this is a city of five hundred thousand people—you had to go to an office of Telepara, the Brazilian phone company, and stand in line for an hour or so to make an international call. This is for the whole city, O.K.? Now, we didn't have any phones at the project. We had radio contact. That happened for a number of years.

HKS: Who did you call on the radio? Somebody in Belém?

JCW: They could call the airport in Belém, or they could call the office in Belém and talk that way.

HKS: Oh, I see.

JCW: So other than that, there wasn't any contact. It was a very isolated situation. We didn't have a private corporate jet or anything like that. We did have the DC-3, but again, that's once a week. Then over time, at some point

Telepara established a telephone link there in Monte Dourado where you could go and wait in line for an hour to use the telephone.

HKS: O.K., that's good.

JCW: Now, they've got telephones in everybody's houses. You can make an international call from the guest house there to the States or anyplace else. So I guess what I'm trying to say about directors, given that kind of a communication situation, if you're Mr. Ludwig, how are you going to talk to your director? If you're a director, how are you going to talk to Mr. Ludwig? How are you going to discuss these things on a day to day basis, probably very important strategic decisions? If you got Mr. Ludwig, who's a very micromanagement kind of guy, and you don't establish a good relationship of trust, and you don't have time to, because you've got such turnover, I think that can create tremendous problems. So anyway, that's my take on director turnover.

HKS: I didn't read many newspaper articles, but in magazines, it was Ludwig's eccentric style that caused the turnover. Of course everything was blamed on him, and as head man, maybe that's fair.

JCW: I think the press is like all of us are sometimes. Simplicity is a lot easier to deal with than reality. We just want to look for one root cause for everything, and I think that we know that life's a little more complex than that. One can certainly blame Ludwig's eccentric style. That might have been a necessary condition for the firing, but it might not have been sufficient.

HKS: I asked Clayton how he survived for all those years. He said, because he wasn't afraid of being fired. He said, if you stopped worrying about that, and you just had to do—even if you disobeyed an order—you had to do what you thought was right. But if you got to second guessing yourself, you would wind up losing the job. That was his answer why he lasted longer than all the thirty-four predecessors put together, apparently.

JCW: I don't know.

HKS: Of course, communications was better by the time Clayton got in charge, too.

JCW: It was a little better. When Clayton was in charge, we might have had one telephone in Monte Dourado. We didn't have much more than that. It was still pretty weak. Who knows what it is between individuals, the rapport that's established? It's hard to say.

HKS: I haven't kept score, but maybe only twice, in all the times you've mentioned Ludwig, you said Mr. Ludwig, and the deferential statements by Clayton, and by Bob, and yourself toward that guy—it's intriguing. You always call him Mr.; he was a special person in your mind somehow.

JCW: Yes.

HKS: And yet, it wasn't called Ludwigandia, like Fordlandia. He didn't have that kind of vanity, did he?

JCW: No, he didn't.

HKS: He didn't put his name on it.

JCW: No, he didn't. He was a very focused individual, for all the good and the bad that is, being focused. It depends on what you're focused on, I guess. But vanity was not one of his focuses. In fact, if we think about how the Jari project or Mr. Ludwig managed the press over the years, that might have been one of his weaknesses. Maybe it would have helped if he had had a little more vanity, so that he would have been a little more worried about how he projected his image. I know a reaction I would get if I were in a conference or something and said, I work for the Jari project, or from a journalist, let's say, Oh, that's very secretive. You people don't share anything, you know. You don't want to let anybody know what's going on in there. Speaking from the people I was around, I don't think that was the purpose. I just don't think Mr. Ludwig felt like it was anybody's business what he did there. It was his money, it was his dime, he was stepping up to the counter, he could spend his dime the way he wanted to, and it wasn't anybody's business. It wasn't that he was really trying to hide anything, as much as he said, Well, what's it to you, you know? This is not your project. I own it. I can do what I want to.

HKS: Sure.

JCW: I think maybe back then it wasn't quite as naive as it seems today. Particularly with respect to the political changes that took place in Brazil. With that *abertura*, and some of what I call just plain demagoguery that happens in a populous type democracy. When populism became an important factor, I think that hurt and caught the Jari project blindsided. So maybe Mr. Ludwig should have been a little more vain.

## Planting Stock vs. Container Grown

HKS: Could be. I read through your articles, and I jotted some things down that struck me as significant. Remembering what Clayton said, how he dealt with some of the issues. Pine planting stock versus container grown. Clayton talks about how he went up to Jacksonville, Florida, to St. Regis, and designed the little carton to grow seedlings.

JCW: The little pots.

HKS: He went through that whole thing; it was important to him. I mean, he told the story.

JCW: Sure.

HKS: You weren't as influenced by that as I was. Tell me, for the record here, about bare rooted planting stock versus container grown.

JCW: We felt like we were getting better survival with the container stuff. We felt we needed that for the pine, given our long dry season, and so we never really looked at bare rooted like they did in southern Brazil and we do here in the U.S. Since labor costs were a lot cheaper, it wasn't a big problem. We don't go to container in the U.S. for two reasons. One is it costs more to produce container stuff, until maybe now with the smaller containers you can do better. The other reason is it's more costly to plant. But we had fairly cheap labor, so labor cost was not as much of an issue, and we felt survival was a lot more important. They're not planting pine today in Jari; they're planting a little bit. And I can't recall if they're bare root planting or if they're using small plastic bullets or not.

## How Much to Grow

HKS: I copied this quote down. "A common directive to managers during this period was the need to develop this resource potential at a fast pace." That's Ludwig's tactic.

JCW: Yes.

HKS: "The desired rate was often made explicit by the setting of planting or future volume targets." That doesn't strike me as profound now as it did when I copied it, but then I've been listening to you for several hours now. You have to have a target. And the mill itself was a target, its capacity, I suppose.

JCW: I think the first targets, when the mill was still just a dream out there, was to say, Well, we're going to plant "x" number of acres. I don't know how that was derived. I think one year we planted sixteen thousand hectares. That might have been the maximum we ever planted. That's a tremendous amount to plant for an organization in one place. That was intended to be the first driver. Then the next driver said, Well, we want to start a mill, in such and such time. Now what do we have to plant to get that volume?

HKS: Would it have been irrational to have had excess production, excess to what the mill could accept? Is there an export market for that? Or would the mill be actually the limiting factor both ways?

JCW: I see what you're saying. I think Clayton sort of started this paradigm, and several of us continued it. I don't know that I would hold as strongly to it now as we did then. Unlike some of the temperate species, where you could plant the trees and you could hold the volume on a stump, with gmelina, you couldn't hold it on the stump, because it would stagnate or it would die or something. Therefore you didn't want to plant more than you needed, and you needed to have fairly fine tuning on that. If I were to redo this thing, I wouldn't accept that paradigm. I think you can hold it longer. Now you can't hold it for ten years, but I think you've got a little more window to plant excess.

HKS: Was there a plan for a second pulpmill and all that? I mean, if everything had gone well and the hydroelectric plant—none of this happened, but...

JCW: I never knew what was on the drawing board completely. There might have been that. One of things was the hydroelectric, of course. The original idea, from what I could gather, was not the pulpmill. It was more solid wood products. I think I mentioned this in my Puerto Rican paper—the first idea was to plant gmelina, grow it on about a ten- or twelve-year rotation for a white wood veneer, and use thinnings perhaps for pulp and have a pulpmill. Some of the major things was to produce a white wood veneer. That's why Weyerhaeuser was down there.

HKS: I see.

JCW: We had a regime at one time for pine—I want to say thinning it at six or seven years of age, thinning it again at twelve, and maybe clearcutting it at seventeen years of age. This is a real early model, back in 1973 when we had the first plan. Originally it was solid wood products with pulp as an aside, and paper. As time went on, pulp and paper became the driver. Now why did that happen? Partly because we saw that gmelina, without some further genetic refinements, particularly on poorer soils, was not a very straight tree. Gmelina has the characteristic; the poorer the soil, the more the tree looks like an apple tree.

HKS: I've seen some photographs of trucks with this twisted stuff on them. That was gmelina logs, probably.

JCW: Yes. On a very good site the thing grows straight, tall. That's probably where the paradigm and the stagnation came in, because I do believe it's true that if you were trying to grow gmelina for solid wood products on a good site, you need to come in there and thin it and keep that growth up, because it will—I don't know if stagnate's the right word—it will lose its ability to respond. It's a fairly short-lived tree, like cottonwood, you might say, but shorter-lived than cottonwood, and so you lose that responsiveness to keep it on its growth trajectory of big diameters. Of course there's where *ceratocystis*, the fungus disease, comes in because that's where we started the pruning, to make clear wood for the veneer option. And you go in there and thin it, and you leave more places for the *ceratocystis* to come in. It tumbled together to where the solid wood product option actually hurt us in terms of *ceratocystis*.

HKS: There were sawmills on site, right? A hardwood and a softwood sawmill?

JCW: No softwood sawmill. There was a hardwood sawmill. Let's talk about two things. Prior to the pulpmill being built, there was a hardwood sawmill, a local one, just for our own consumption. Then we tried to export out of the project. I don't know where it was sold to. Then when the pulpmill was built, there was a sawmill built in conjunction with the pulpmill for sawing native logs. Actually it had two reasons. One was to saw native logs for producing lumber for possible export, but the other was for a chip production. Part of it was to produce chips for fuelwood. So native wood, once the mill was on stream, went in three different directions. Those species that had been identified as pulp species were used for pulping. Those species which didn't have any lumber value went directly to fuel. And those species that had some lumber value were sawed for lumber with the other part of it going to hog fuel, for fuel for the boilers in the pulpmill.

HKS: O.K. Clayton talked about this large sawmill, large in terms of to saw large logs.

JCW: That's what I was talking about.

HKS: It was a battle that he lost, that there is still enough stateside thinking driving the decisions down there that he wound up having to accept this large-saw sawmill.

JCW: The mill was designed, I think, by Simons out on the West Coast. I think what Clayton's referring to was that that mill was really designed more like a west coast sawmill, a softwood sawmill. As most people in the industry know, softwood and hardwood sawmills are totally different animals. A softwood sawmill tends to be a mill that's going for a maximum throughput—

HKS: O.K.

JCW: A hardwood sawmill, it's going for quality, being very careful about how it saws up. I think the battle Clayton is probably referring to is the battle that that thing was designed more for producing chips for the fuel and stuff. It wasn't being run like it was a hardwood sawmill. It was tied to that pulpmill.

At one time, even after that was built, we put some thought into maybe even building another smaller sawmill, or doing something with it to try to capture some of the lumber value. But that's a whole different issue, the whole thing of how to merchandise those trees and find markets for those trees. Those are upland species, and the worldwide markets are accustomed to species along the river in South America, that float, with lower specific gravities. We spent a lot of time trying to market—we had a wood tech lab, did a lot of testing and so forth, and with mixed success trying to market it.

HKS: Let me pick up a point on gmelina and the pruning that made it susceptible to the chancre disease. The pruning was done to create clear wood.

JCW: That's right.

HKS: For a non-pulp market.

JCW: That's right.

HKS: So was it with the idea, at that time, that the thinnings would produce enough pulp?

JCW: If you go back to when that was the paradigm, O.K., of producing lumber and so forth, then, yes, it would. The question at that time in the planning, and one of the things they looked at was, Well, let's see when this pulpmill needs to come on so that we can regulate the thinnings. A lot of wood to generate from thinnings. This is one of these things that was discussed, I think, up in Stamford, Connecticut, as they talked about different things. There was probably a communication problem in that a lot of this facility's planning was being done up there, and yet the data for doing the yield projections—not only the quantitative data but the field operational planning was down at the project, many thousands of miles and not a phone call away. We know there are problems that exist like that even in the States, where people are located just two offices down. So this thing about the solid wood and the thinnings generating the pulp—I don't know how far that got. But I believe it was abandoned fairly quickly, within a year or so. It became more clear that we would have to generate that volume if we wanted a pulpmill to go with more clearcuts.

HKS: I'm pursuing this gmelina business and the pulpwood mentality, when that set in as opposed to having saw logs or veneer blocks, or whatever, and the shifts that it would make in terms of management decisions.

JCW: I don't know when it set in chronologically. I know what kind of caused it is we looked at the yield projections. I want to say it set in in about 1976 when some of this planning was starting to be done, looking at sawed wood. By that time we had some yield projections that were indicating—we had plantations by that time that were planted in 1969, that we had the plots measured over time on some fairly good soils. What we were seeing was that the mean annual increment, the volume productivity, was peaking earlier. When we had thinned some plantations, and they were not responding to the thinning, which was saying that if you wanted to go with a sawed wood cycle, you were going to have to take a hit in terms of total productivity. You were going to have to either have big trees or you could have more volume per acre per year. It was becoming apparent that that was a problem.

We had an experiment, in fact it was the initial experiment that Clayton put in for pine in 1970, where we had thinned that at about age six or seven and the mean annual increment curve. The trees did not respond to the thinning. My working hypothesis is the reason that didn't take place with the pine is that we had so much genetic variation in that initial seed source. Now we didn't have the fertilizer experiments, I don't think, in mid rotation time, but maybe with proper fertilization you could get them to respond more to the thinning. But I primarily attribute the problem of not responding to thinning as genetic in origin. I think that with the right genotype you probably could get a better response.

That sort of laid the ground for some of the difficulties for a solid wood products regime, because if you're going to take a hit on your mean annual increment by thinning, you have a fork in the road. You're going to have to have lower productivity or you can say, Well, look, I'd rather have higher productivity to generate more fiber and go for a pulpmill or something.

HKS: Given that chancre was a problem with gmelina, and the chancre was a problem at least in part because of the pruning, when you went to a pulpwood paradigm, you would stop pruning.

JCW: That's right.

HKS: Didn't that help alleviate the chancre problem?

JCW: That was our hope, that it would reduce the chancre problem. I think it probably did. If we had kept pruning, we would have had even more chancre. But you still had chancre out there, and on top of that, you've got the problem of when you coppice the trees, you're just creating a place for the chancre to survive and to get in there. I think the chancre problems with gmelina probably could be solved over time with genetic testing. But again, do you want to spend your resources doing that, or would you rather go to something like eucalyptus? This is again where research comes into play, and we put out some eucalyptus trials. The company, after Mr. Ludwig's time, put out additional eucalyptus trials, and I think the conclusions they've reached today is to say the better strategy is to go with *Eucalyptus urophylla*, or something else, and don't pursue the gmelina strategy, which to me seems rational at this point.

HKS: There are several hundred species of eucalyptus, right?

JCW: Yes.

HKS: How many are commercially viable?

JCW: I don't know.

HKS: About a dozen? I mean, a small number.

JCW: I'd say a fairly small number.

HKS: But enough is known about eucalyptus that you could make logical choices of what to experiment with.

JCW: The other advantage, from an industrial standpoint, with eucalyptus is that you're not the only one working on it. In other words, you've got a wide variety of experience, albeit not in the same location and on the equator, but you've got a lot of facilities working with eucalyptus.

Aracruz got some of their first start from eucalyptus in Zaire, which is close to the equator. Now I don't know enough about that experience in Zaire with eucalyptus. I just know that from some of the history I've read is that that's sort of where Aracruz got some of their first start, was with some of the eucalyptus in Zaire.

## Leaf Cutter Ant

HKS: The leaf cutter ant comes up a lot in the literature, and you mentioned it, and the use of Mirex or whatever.

JCW: Yes, Mirex.

HKS: It was a real problem? A serious problem?

JCW: It's the kind of problem that's an endemic problem. It's endemic in the same sense that southern pine beetle is endemic to the southeastern United States. The difference is the way you go about controlling it. With the leaf cutter ants you find ways to kill the pest. But you're never going to eliminate it. Almost like fire control, you just put that in the budget as an annual cost. You're never going to eliminate fire, but you're going to put in methods to reduce the incidence of it. It's the same thing with leaf cutter ants. Leaf cutter ants are a problem not just to Jari. They're a problem all over Brazil, in one form or another. They just learn to live with it.

HKS: I was touring in Costa Rica at the OTS operation at La Selva. For us gringos, the leaf cutter ant was something wonderful. We'd stop and look, we were in awe.

JCW: They really are neat.

HKS: They're marvelous things. I didn't realize they were a problem until I started reading about the Jari experience. Do they kill the tree? Or just defoliate?

JCW: They can kill young trees, I guess they can even kill older trees if you let them defoliate that same tree repeatedly. But generally the biggest problem is that they reduce the growth quite a bit, because you have less photosynthesis taking place.

HKS: The leaf cutter ant is an issue, in a plantation situation; it wouldn't be under so-called normal, uneven age stands, because you have this concentration of small trees.

JCW: I'm not sure I understand. Leaf cutter ants attack big and small trees. In other words, in the native forests, they go up big trees and cut them. It's just that it's not a problem because you're not trying to maximize wood production. The native forest there is probably not producing in terms of merchantable volume more than three cubic meters per hectare per year. Whereas in Brazil with eucalyptus you're trying to grow thirty cubic meters. So you're trying to grow ten times as much fiber per hectare per year than you are in the native forest. If all you're trying to do is produce three cubic meters, or four cubic meters, leaf cutter ants aren't a problem because you're just taking that into account. But if you're trying to maintain these high yields and these insects are coming in there cutting down your photosynthetic factory, then they become a real problem.

HKS: Genetic selection would work for diseases, but not for insects.

JCW: I'd hate to say never, but it would be very difficult, it seems to me.

HKS: I guess it would taste bad or something. I'm not sure why insects wouldn't do that.

JCW: The thing that's always interesting to me about diseases and insects, and we struggled with this in the Southeast with *Cronartium fusiforme* on loblolly pine. We talk about getting genetic resistance. But sometimes when we say it, we forget that we've got two organisms here, we don't have one. And both those organisms are capable of adapting. They're playing a game here, because once you move the genetic genotype on your primary species being attacked, you cause some changes in the genotypes of the pathogen.

HKS: Sure.

JCW: That's why I say when you think research in the tropics, just like in agricultural crops, you basically are entering a game of one genotype against another genotype, the pathogen against the host. We are trying to manipulate with our selection and adding to our arsenal ways to counteract that. We've had a lot of success in domestic agriculture with that strategy, and I assume we're going to have a lot of success with that in industrial forestry.

HKS: Was the leaf cutter ant a problem with pines, or only the broadleaf.

JCW: Both.

HKS: What diseases bothered the pine?

JCW: We didn't really have any diseases bothering pines. I think there was a needle cast disease we would have, but it was the kind of thing that, like most needle cast diseases, we would just see some brown foliage, but it would go away. You couldn't come up with any economical way that it would significantly increase your volume production by getting rid of it. It wasn't really a problem, so I would have to say in our situation, there were no significant diseases with pine.

HKS: The insecticide you used, Mirex. That was the most effective?

JCW: We used three different approaches. One was the Mirex, which was the bait. It's the same thing that we use for fire ants here in the U.S., or very similar. With that approach you would broadcast it. We put it in plastic sacks to keep it dry, because if it rained the humidity would ruin the chemical. That was one approach. The other approach was using methyl bromide at the time, and you would stick a hose down into the ant hole and you'd basically fumigate the ant hole. The other approach was another fumigation method that is still used in Brazil. There's a chemical in the sprayer which hits a hot shield, which atomizes the chemical. Then you spray that down into the hole to fumigate the hole. So that's the third method. We used all three. We had a forest entomologist on staff, and most of his time was spent with the leaf cutter ants problem and improving methods of treatment.

HKS: One would think there'd be an extensive literature on that, because if it's so widespread, everyone has this same problem.

JCW: I have a book, in fact, called *As Souvas* which is the name for leaf cutter ants in Portuguese. There is an extensive literature in Brazil on the leaf cutter ant. But it's a little bit like there's an extensive literature on psoriasis and asthma. There's no single, complete solution to it.

## Work and Workers

HKS: I don't want to sound like a journalist if I ask this question.

JCW: That's O.K., go ahead.

HKS: Worker safety. People are handling toxic materials. You're putting toxins into the environment, in terms of our standards here in the States, EPA's requirements and federal statutes, state laws and so forth. How much of an issue is that when you work in Brazil? Are you accused of not having the workers properly clothed?

JCW: We operated according to the standards of Brazilian labor laws. We worked within the culture that we found ourselves.

HKS: I understand you can't use DDT in the States, but it's used in other parts of the world.

JCW: It might be. We didn't use DDT.

HKS: That's an example. It's something that's illegal here but it's not illegal someplace else.

JCW: I know. Different cultures have different standards.

HKS: Since Jari was seen, at least in the later years, as imperialism or as not a good thing for Brazil, because it wasn't Brazilian, were these sorts of things ever an issue, the double standards? Your answer is no.

JCW: We tried to be consistent with the labor standards. There were Brazilian authorities that would check up on us. We followed all the same labor laws, everything else.

HKS: I can remember when I was a forestry student, taking silviculture; we went out on a field trip. This is in the late '50s. We used 2-4D and 245-T without any protection, any warnings.

JCW: Yes.

HKS: We didn't wash our hands afterwards—

JCW: No.

HKS: Nothing. Zip. You don't do that anymore.

JCW: No, you don't. I've had the same experience, of using 2-4D and 245-T without the same level of safety precautions that we use today. It's not for me to judge. I'm not talking about Jari or anything, but it's not for me to judge what is the appropriate level of rigor, because I think it has somewhat to do with people's tolerance for risk and uncertainty. Some people, we know, are very risk averse.

HKS: Sure.

JCW: They'll avoid any situation at all. That's why they use organic things. But coming back to Jari, we basically followed the legal standards of the society.

HKS: They were well-articulated standards? They weren't vague, in terms of whether you had to wear protective clothing or not?

JCW: Yes.

HKS: It was spelled out.

JCW: In general you'd say that there weren't many standards.

HKS: Hard hats weren't required, like OSHA requirements.

JCW: No.

HKS: Nothing like that.

JCW: A lot of us didn't use them. We had some accidents, but relatively few accidents for the number of people we had working—and I can't back this up, because I don't have any statistics. My perception is, incident rate per number of workers was probably not a lot higher than it would be in the States, because we had such a sheer number of workers. But where our biggest safety hazards were, to me, is in vehicles. We had dirt roads, dusty in the dry season, which would go from roughly August through December. You had these dusty roads, and I used to think that a lot of Brazilians felt like they were Emerson Fittipaldi, the race car driver. Of course Rio has a big reputation for bad driving. I think exceeding the speed for the conditions was our biggest risk. In fact, one of the workers I talked about that used to work for me, Cristovam (it was after he stopped working for me and while he worked for a logging contractor), was killed. He was killed in a head-on crash with a pulpwood truck. I warned Cristovam several times about his speed when he was driving a pickup, because people had seen him speeding, and I'm sure it was excessive speed. The contract laborers would be in the back of a truck, which of course you couldn't do in the States. You'd have to have a bus. I guess we should have had buses, or Brazilian society should require buses. If one of those things turned over, people are going to die, or somebody's going to get hurt seriously. I know it happened one time.

HKS: Clayton, as I recall, mentioned the difficulties in keeping certain employees who were technically acceptable. They philosophically didn't want to work in a place that seemed to be exploiting local labor, who didn't provide adequate medical care for the Brazilians. This is Americans working, basically.

JCW: Yes.

HKS: He didn't expand on this. It was almost an aside. I didn't know if that was a major issue, or he just remembered a couple of guys who really thought, Ludwig has billions. Why aren't we building more hospitals?

JCW: I never discussed that with any of my peers. My training at Yale was a master's degree in forestry and economics, and the courses I took at Yale in economics were primarily in economic development. I took a macro and a micro course, but also took mainly some courses in economic development. At that time one of the models was the model for economic development in a labor-surplus economy. Basically my attitude was to look for the appropriate technology for the cost of labor and the availability of the labor in the society. The appropriate technology is not necessarily the most capital intensive technology available. From a western European perspective, you could come in there and you could say, Well, look at Jari. They're exploiting these laborers. These people should have air conditioned cabs in their vehicles, O.K.? My perspective was, Well, if we do that, then we are going to get rid of laborers, because capital is going to be substituting for laborers. Where are these people going to work? Is it better for these people not to have a job, to go back to northeastern Brazil and maybe starve or work at subsistence wage and die of some parasite? Or if it's their free choice, because these people were not slave labor, if it's their free choice to come and work under these conditions, save some money, go back—later some of them brought their families and became permanent workers—who am I to judge? I subscribe to appropriate technology, labor-surplus economy type model.

We were talking about plantation management, giving a contrast to Jari in the early years versus what was done in southern Brazil by some other firms. One of the interesting things to me was we adapted what I considered an appropriate technology for silviculture and harvesting, which was not as capital intensive. That is, we didn't go for feller bunchers and all the rest of the stuff. As you've read and seen, we used a big stick harvester, which was used in the South up until the early '60s, and which is a fairly labor intensive method of harvesting. Now it's not the most safe thing to do, or certainly we can find a safer way of harvesting, but it gave people jobs. It was the lowest cost method of harvesting, given our labor rates and the economy as a whole and a perfectly competitive society.

As an economist, I believe that in general the best estimate of the true cost of resources to a society is the prices observed in a free market. Now granted, we can think of externalities and so forth, which create exceptions to that, but in general our best estimate is the market price of things. If the labor cost is thus and so, and the capital cost is thus and so, we ought to use the appropriate technology based on those costs.

For various reasons the firms in southern Brazil decided to go to a more capital intensive approach. They have feller bunchers and all sorts of stuff. We can talk about some of the reasons that might be. But one reason may be because that was what they were familiar with. They had more exposure to labor problems. In other words, it was kind of a hidden cost of labor.

In later years Jari has moved toward a more capital-intensive approach because of the costs of the infrastructure. If Jari's going to have to furnish the schools and other infrastructure (as opposed to the government), then there's this hidden cost of labor, which is saying, Well, you need to get more intensive in your harvesting systems. And that's what Jari is doing, for better or worse. Now one can argue from a Brazilian perspective as a whole, maybe the government ought to provide the infrastructure so the labor costs wouldn't be so high, and so you could have more people employed gainfully and make a living wage.

HKS: In the States, I've been told, one of the things that wiped out the naval stores industry was a shift in the federal laws, that workmen's comp is now required. It increased the cost of labor.

JCW: That's right.

HKS: They stopped producing it by that method. Then all those jobs were lost.

JCW: Yes.

HKS: I don't know if it's a philosophical question or not. Whenever the debate in Congress is about raising the minimum wage the same kinds of issues come up. You make the jobs more expensive, and McDonald's will start automating certain things.

JCW: Right. I don't know if it's philosophical. It's just a fact of life that if you're a producer, an entrepreneur or whatever, and the cost of any resource goes up, then you're going to look for substitutes for that resource. That's the issue. Look at safety. If Jari was to say, Well, the safety standards of Brazil are not good enough for us. What is the reaction? We are going to give everybody steel-toed boots, we're going to do all this stuff, we're going to use buses, but you have to say that increases the cost of labor as an input. Which then means you need to go to a more capital-intensive system. I think it becomes a philosophical and income distribution issue to say: Is it better to do that and hire fewer laborers, or is it better to accept the standards of society and hire more laborers? I don't know the answer. That is a philosophical issue.

I don't know the dates. There was a company newspaper called the *Jarilino*. Newspaper may not be right—newsletter would be appropriate, I guess. This one's July '78. We were talking about safety equipment and that sort of thing. It's in Portuguese, but it says, Acceptance of safety equipment by personnel. This is an article that's talking about the issue of how do you get workers to accept the use of safety equipment? I know after the mill got started, there was somebody responsible for safety. This sort of article kind of gives you a feeling for where we were at. There weren't Brazilian laws that said you had to use this safety equipment in rural settings. But we were trying to get people to just start thinking about the use of safety equipment, why you should use safety equipment. That's what this article in the newsletter was about.

One of the unique things in our situation at Jari was the weather, being so hot. Much of the safety equipment that I've seen for woods workers is designed for the northern hemisphere, Europe, for example, where you have fairly a temperate climate, cool weather. If you wear that safety equipment on the equator, you'll wind up with dehydration, for woods workers, at least. For example, chainsaw protective coverall leggings and all this sort of thing. One of the biggest safety issues I've felt personally, when I saw workers out in the field, manual laborers, was basically just having good shoes. Some of these folks would wear sandals. One of the common things they were issued was kind of rubber boots, they were good for keeping your feet dry. You know, this is a common Brazilian type footwear, but they didn't give you much protection if an ax should hit your foot. I think footwear was probably one of the biggest pieces of safety equipment that we tried to push and get people to wear, and gradually get better footwear for folks, starting from some of these people were just wearing sandals, to get them to wear at least tennis shoes and covered type footgear.

HKS: Your son Noah was born at Jari. There was a physician on staff. You had a hospital. And all this was free—part of your compensation, as it were.

JCW: Yes.

HKS: Clayton tells the story about, when he first got there, his son cut himself. And Clayton himself sewed him up, the stitches. It progressed substantially, as fast as that infrastructure could be built.

JCW: Yes.

HKS: What would have happened if you had a major illness or injury and you had to be evacuated to a First World hospital, as it were?

JCW: I don't have exact chronology, but basically it evolved from having one plane flight a week to having one plane flight every day. And I think there might have even have been a period, there might have been two flights on one day. The company at one point had two DC-3s. At another point they had one DC-3 and also a Fokker turboprop passenger plane. And they also had two Islanders, which were smaller, about ten-passenger, twin engine planes. So if you had an emergency, there was usually a plane available, particularly an Islander, on site, so you could be flown to Belém, which was about a two-hour plane trip. Where you'd go for emergency treatment was to Belém.

HKS: But that would be free?

JCW: The plane trip would be free, right. Now with respect to the hospitalization in Belém, I'm not sure how that worked. Brazil has socialized medicine, but like socialized medicine in other countries, there's sort of a two-tier system. You have socialized medicine, which is free, but there's also private practices that doctors have. The comments by Brazilians and others was that if you really want top class care, you really would go through the private network. I don't know how it is today.

HKS: There wasn't such a thing as a company health plan, like Blue Cross Blue Shield, like you have here at major corporations in the States, where you knew that a certain level of health care would be provided.

JCW: I never had to use the company health plan.

HKS: As the community developed, you had more and more spouses, more and more children, and those issues became more important.

JCW: Yes. I was never in a situation where I had to use medical care in Belém, for example, and certainly not in the United States. So those issues never came up with me. I used the same health system that everybody else used, the Brazilians and everybody else. I do recall a couple of people who had health problems like malaria and some things. It might have been Don Haight; he went to the Aschner Clinic in New Orleans, and I think Mr. Ludwig took a very active interest in that. I did get the same sense that he cared a lot about people's health and took an active interest in the upper management's situation. National Bulk Carriers, which was Ludwig's company, had a health plan, but I never made use of it, so I don't know what it was.

HKS: I'll have to talk to somebody who got sick. It's part of the philosophy of developing the town.

JCW: Yes.

HKS: I asked you earlier about whether or not you could buy liquor, or whatever, at Monte Dourado. More to find out if Ludwig imposed his morality, his sense of values on the development of the town. And you said no, not as far as you could tell. On several occasions Clayton characterized Ludwig as having a seaman's mentality. You had the ship's captain at sea, and then the captain was in charge. Clayton didn't say this, but on board ship you don't have any liquor. You can't afford people getting drunk and so forth.

JCW: That may have been something that evolved. I think when the project first started there weren't families out there. Probably that was initially a very captain's-seamen's mentality, you know, in which I would probably agree with, to a large extent. You don't need people not staying on task. Then as families came in, it probably transitioned into a little different thing. Then as the town gradually became more open, and more families and so forth, it evolved where some of those strictures were broken down, just like in any normal setting.

HKS: There's the Brazilian shantytown, I guess it's fair to call it, across the river.

JCW: Yes, the Beiradao, they called it.

HKS: How did you get there?

JCW: You'd take a boat across the river.

HKS: This boat, was it a commercial boat?

JCW: Private boats.

HKS: You'd pay somebody a peso or whatever it is and you'd go across.

JCW: I was in Thailand one time, and it's very similar to Thailand, river taxis.

HKS: In a sense, the shantytown was market driven.

JCW: Very. Completely market driven. The shantytown was in the federal territory of Amapá, not the State of Pará, and I don't know the different legal issues. We were talking about trying to develop some plantations on that side for many years. Whenever we talked about that, Mr. Ludwig would say, No, we don't want to develop over there right now. It might have had to do with the fact that it was a federal territory. He didn't want to be in two different government areas. The disadvantage for us from a forestry standpoint was that it was sitting right across the river from where the pulpmill was going to be sited. We wanted to keep our transport costs low.

HKS: Sure.

JCW: And subsequently that was developed. Probably the best production plantations right now that Jari has, the eucalyptus plantations.

HKS: By federal territory, that means it's less developed infrastructure.

JCW: Yes.

HKS: Like Alaska was and Hawaii.

JCW: Exactly. Exactly.

HKS: At some point it would be able to be a state on its own right.

JCW: Yes. If they want, you know, if the federal government wants to and if it wants to.

HKS: I see.

JCW: At various times, during transitions of people, there were company files that gave what the history of plantations were. We have also talked about the management information system. That's contained in file folders at Jari, a record of what was done. We put out a commemorative report which essentially was kind of a closing summary report on the status of the project in the end of 1981. That's a document that exists on the project, of the way things stood.

About June of '83, many of us who had worked for the new company left. We put out a series of five or six booklets about an inch thick each, which basically were operating procedures manuals. There were two of them for harvesting, there was one of them, maybe two, for management, and there were two of them for planting and inventory and control, which was my section. That was a passing over of what current procedures were, and for the new company to have some documentation. Of course, they also had Brazilians in management positions that were continuing on. But it was sort of a transition, some hard copy documents to carry on the transition.

# Labor Question

JCW: I have a concern that there may be a tendency in Brazil, from my perspective, to overmechanize. It gets back to this issue of labor surplus. I have a soft spot in the sense that I hate to see people unemployed, and to see them unemployed because of an overcapitalized harvesting system. So that's a concern for me, and I don't think Jari is the only place. On the other hand, to give those companies their due, in large part, they're just responding to higher hidden costs of labor, for example, bad publicity. In other words, if you get a newspaper article saying, Well, you're exploiting workers, I can't put dollars and cents, but that's a cost of having those workers.

HKS: That's in one of your papers, that labor-intensive can be seen as labor exploitive.

JCW: Exploitive. that's right.

HKS: And so if you mechanize and reduce the numbers of people, you increase the salaries of those who remain, because they're more technically skilled?

JCW: Sure. You have a lot more investment in those people, because you're all of a sudden bringing labor up who's highly skilled, who has the ability to go someplace else and get a job in that economy.

HKS: So the ones who remain on the payroll are better off.

JCW: They're going to be better compensated. They have better safety equipment and all this other stuff. But you don't have very many workers. In the back of my mind, there's this guy who's doing great, family looks healthy. But what about all the families of the other people you're not hiring, that you don't see, you know, that aren't around you? The ones that are still in northeast Brazil without a job. That's the thing that goes around in my mind, when I think of overcapitalized systems. Most of the people who would listen to this can think of a lot of folks in the world who are underemployed because they don't have the skills that match the technology that's out there.

HKS: When I read about Monte Dourado, over the years, the population bumps up and down. Twelve-five maybe was the highest figure I saw, and now it's down to eight thousand. Is that because there's less development? I'm not sure I have the numbers right.

JCW: I'm not sure either. There were three different populations that I would, if I were doing a census, try to think of. One population would be the people who were direct company employees. The other would be the population of contract laborers. And the third would be the population of people that lived across the river, who didn't work for the company but were supported by the company because people would spend their money.

HKS: Service industries.

JCW: Service industries, that's the word. I want to say the population today is probably thirty-five, forty thousand people.

HKS: Is that right?

JCW: When people would come visit us, they would always ask these kinds of questions about the population. We had some numbers about how many people lived in those different categories, I believe. I don't know about the service industry category, but at least with respect to contract laborers and company employees we have pretty good statistics for those. Brazil probably has at least some census of that area. The federal territory has a census probably, and Pará has a census.

HKS: You said off tape that you had to get permission to add a porch to your house. And you had to pay for that. Are there houses in town that people construct? They're not actually company built houses?

JCW: When we were there, no. The rent was very minimal. In fact, for some time it was probably free. But then as time went on there, they put a rent on it. But they were company houses, and the idea was not to allow people just to start adding things on to these company assets. The idea was not to prohibit you from doing something, but to make sure that whatever you did followed some reasonable architectural guidelines, and engineering guidelines, for structural integrity to do it.

HKS: How did you accommodate new labor? Was there a barracks situation?

JCW: There were barracks out in the field at different places.

HKS: O.K.

JCW: Permanent barracks. There was a village built up on the plateau near the airport. There was another one constructed out near the nursery in São Miguel. There was one constructed down near the kaolin plant. The idea behind this was to allow laborers to bring their families in there, and settle, and have their offspring become future laborers for the project. We really wanted to establish a town there. This was not a kind of a cut out and get out kind of mentality.

HKS: O.K.

JCW: This was to try to establish a permanent enterprise. A Brazilian mentioned to me one time that for somebody coming from southern Brazil—it was just about as much of a shock coming to where they were as it was for somebody from the United States coming or living there in the middle of nowhere, so to speak. Our concerns were to try to develop a town, have schools and things, where people would want to raise their families. The company would have a labor pool, and they would have a pool for having service industries and all the things that comes with a city. I think one of the disappointments of Mr. Ludwig and the group was that the government in many ways didn't want to accept some of the responsibility in helping us develop a permanent town there. In other words, Mr. Ludwig had to carry on his shoulders many of the infrastructure things that one would associate with government providing services because of the taxes that industry and people pay. That continues to be a problem, even for the new company.

HKS: Clayton mentioned how Jari even had advisers to teach the locals how to have gardens, what vegetables would grow well there.

JCW: There was a fellow named Anel Guevara, who was a native of Panama. Anel had several jobs over the period of time. There were some outlying villages. Now a village here was about two houses, or three or four houses, from the old Brazil nut collection days. Anel would go out and work with these people about how to grow, how to raise vegetables and produce to sell in Monte Dourado, and we set up a market in Monte Dourado. So Anel was actively involved with that. A lawyer by the name of Ana Moura, that was part of her job, although she was also responsible for working with contract laborers and doing a lot of that. We had one or two social workers, who would also work with the families in the Silvavillas, for example. So that was some of the things we were trying to do.

HKS: In the early photographs, before any of the street trees grew, it looks like it was laid out by a military regime. But you look at the photos now, at least from the air, there are trees all around.

JCW: Well, things grow pretty fast. That's an advantage. The photoperiod at the equator was an important factor. That really wasn't a problem in terms of fruits, because they were tropical fruits. Now granted you couldn't grow apples and things like that. With vegetables, the photoperiod and the heat gave some problems for some of what we would like to have. Americans at least were accustomed to green vegetables. So there was a difficulty in getting green vegetables. There are some lettuce varieties that you can produce there. I would say that the biggest lack for people—this is true of Belém, at the big markets—is green vegetables. The Japanese Brazilians were really good at growing things, green vegetables and so forth. But the photoperiod limited what you could grow. You couldn't grow broccoli, for example, and some other things. And some of us tried a lot of different things in our own home gardens, but couldn't do it.

HKS: But you never felt you weren't getting a nutritious diet?

JCW: Oh, no, no, no. Beans and rice, which is a Brazilian diet, is almost a complete protein diet. In fact, I think we probably ate a lot better there than we eat here in the States.

HKS: There was no Dunkin Donuts there.

JCW: Like right now I'm eating a Dunkin Donut. In terms of the infrastructure, I was about to say something about Ana Moura. We had different companies that provided contract laborers. We needed to understand what their cost structure was, and we needed to understand if they were following the legal requirements of Brazilian law. We established a contract labor company separate from Jari, and it was called SASI. I can't remember what that stands for. But Ana Moura is a lawyer, and she was the one that ran that company, and along with, I think Almir Bastos was his name. That was our way of understanding about the legal and the physical requirements, logistical requirements of getting contract labor. It was sort of our benchmark. We did a similar sort of thing, and actually it was SASI that did it, when we got into harvesting. We eventually had other contractors in charge of harvesting, just like you do in the States, contract harvesters. But we had our own company and crews, plus SASI did its own contract harvesting, so that allowed us in a way to understand what was going on and to try to provide some guidance, some technology transfer to the other harvesting contractors.

HKS: For harvesting contractors to come in, would they have to bring their own equipment? It would seem to me that's pretty expensive.

JCW: I don't know exactly how we did it, but we financed them in terms of the big stick harvesters. One of the good things about these big stick harvesters, of course, is they're very low cost. Now all the harvesting for the native forests, which required skidders and that sort of thing, was company harvesting. The contract harvesting I'm talking about had to do with using the big stick harvesters, which was a labor intensive job, which basically involved a Mercedes truck, which was built in Brazil, a kind of standard truck in Brazil. At first we imported some of these big stick harvesters from the States, and then I think gradually we might have manufactured some of our own. I can't remember.

HKS: We have a community from ten to twenty to thirty or more thousand people. How many Americans were there? A hundred? Two hundred? I don't want an accurate count, but what order of magnitude would you guess?

JCW: Of course, one thing we're talking about here is a big time span.

HKS: Sure.

JCW: So any answer I would give would be wrong for one particular time span. So let me describe the different phases of the project. There was sort of the first phase, when people came in and established the town there. That was before I got there. The next phase would have been the phase when you geared up for planting and plantation management on a bigger scale, and planting tens of thousands of acres of plantations. And getting the road infrastructure in there. The next phase was gearing up for when the pulpmill would arrive, the civil engineering phase. The next phase was once the pulpmill had arrived, the start up phase of the pulpmill. And then the last phase was the running of the pulpmill on an operational basis, so I don't know, maybe I've described six phases there. It wasn't just Americans but a lot of expatriates. We had people from England, we had people from Angola, Mozambique. We had a lot from other places in South America besides Brazil. It's hard for me to come up with a number. I want to say a hundred.

HKS: That's really what I'm after.

JCW: The ballpark estimate. During the pulpmill startup, for about a year, there was a phase when we had a lot of people from Finland come there. *Jari: Seventy Years of History* by Cristovao Lins here. It's in Portuguese. It gives a list of all the different nationalities that were there. That's pretty accurate; Indonesian, Italian, he's got Indiana. I don't know if that's a state or.... [laughs] Oh, he must mean Indians. I guess that's Portuguese for Indian.

HKS: You either spoke English or Portuguese.

JCW: Yes. Like the Finns, you know, when they were there, they either spoke English or Finnish. Not many of them learned Portuguese. They didn't have any reason to learn it. It depended on who you were with. I would speak English at home, and I would speak English if I was just around Americans without a Brazilian present. But Portuguese is what I spoke.

HKS: Did your wife speak Portuguese?

JCW: She spoke enough to get done what she had to get done.

HKS: Shopping and...

JCW: Shopping and that sort of thing, yeah.

HKS: Did she work?

JCW: In the home. There was an international school there for a period of time, for expatriates. It was an English speaking school. And she worked in the library there, to help set up the library. I'm not even sure if she got paid for it. She might have volunteered.

HKS: I'm way ahead of you on the donuts. That's the advantage I have of asking the questions. Anything more to add about life in Jari, before we get back to some of the technical questions?

## Importing Technology

JCW: Not really. Earlier we were talking about any generalizations I could make. I talked about generalizations with respect to plantation forestry. Another generalization I think you can make is a generalization that's not new, certainly not new to warfare, is logistics. I think you have to pay careful attention to logistics, and I think part of the appropriate technology is to not have technology which you can't support logistically. One of the concerns we had in developing our harvesting systems and any kind of engineering thing was the fact that getting parts was not always easy. If something broke down, you needed something that you could get the parts. Even better, you needed something that wouldn't break down, so you wouldn't need to get the parts because it would be very reliable. That's of course something true of anything in a remote location.

HKS: When you ordered equipment, did you tend to order from the States, or did you order from suppliers in Brazil?

JCW: If you could find it in Brazil, you wanted to get it in Brazil. But oftentimes Brazil didn't have the equipment you were looking for. Caterpillar tractors, for example. They were ordered from the States, although eventually I guess we got them from Brazil. I don't know exactly how the Caterpillar tractor distributors worked, where they actually were manufactured, but I recall there were a lot of difficulties in the early years, in trying to get some specialized logging equipment in to test things, because you had to go through a pretty stringent process with the government. Is this machine made in Brazil, or something very similar to this machine? And then you'd go through all the bureaucratic hassles and say, No, this is not exactly the machine. The people doing all the haggling and arguing, of course, were people who were not directly going to use the machine. We tried to find Brazilian equipment where we could. I know for a long time, one of the problems was computers. We couldn't bring PCs in, because Brazil at that time was trying to develop their own PC industry.

HKS: O.K.

JCW: We wanted to use the latest technology from a forest analytical perspective. We really couldn't use it, because we were barred by the government from bringing that kind of stuff in. They've either caught up or they've gradually abandoned some of that, and there's a lot more free market availability of these things. They have also bought into the global economy like many of us. But we were living in a period with tariffs and all that sort of thing, not having the availability of some types of equipment.

HKS: But when communications improved, so you could call directly from Jari, you could call the States and they'd air ship a thing down and you'd have it in a week or a few days.

JCW: If you got the customs clearance.

HKS: I see.

JCW: You might have the communication links set up, and you might have the transport link set up, but you've still got the bureaucratic governmental barriers.

HKS: Your office in Belém, one of the things that they did for you was deal with the bureaucracy.

JCW: Exactly. Exactly.

HKS: Was that office just a few people, or was it a large office?

JCW: It's hard to remember that one. Fifty people, maybe. I may be overstating.

HKS: But maybe the bureaucrats—

JCW: The ones that really dealt with that kind of issue, there were probably no more than four or five people that dealt with expediting things. *Despachantes*, I think, is the Portuguese word for it, people who expedite different things. Back to the generalizations. I think you have to pay careful attention to logistics. Another thing, I think the government and you have to have a clear understanding of what they're going to furnish and what you're going to furnish. Now that's easier, obviously, said than done. We can't even do that in the United States. But it would certainly help a lot of situations. If you don't have a clear understanding, you should certainly hedge your bets and have options in case the government should renege on some of their promises.

HKS: Was bribery an issue, to get things done? As it is, say, in Indonesia.

JCW: I didn't have any firsthand knowledge of it.

HKS: I'm not trying to get you into trouble.

JCW: No, I understand. I didn't have any firsthand knowledge of it, but my working hypothesis, you might say, was that there were times that—it wouldn't surprise me.

HKS: A guy like Bob Gilvary who had to bring in a lot of equipment must have needed shortcuts from time to time.

JCW: It wouldn't surprise me.

HKS: It wouldn't surprise you.

JCW: No.

## Jari and Economic Models

HKS: A question I should have asked earlier when you were talking about your yield tables. The mathematics of tropical plantations, the formulas and so forth, is it the same as temperate? There's no different kinds of mathematical derivations of biological properties? You had to come up with new kinds of equations. The fundamental things you have in American forestry textbooks—

JCW: You can use those.

HKS: But you had to measure different things, and the numbers are smaller, I mean, the rotations were shorter and all the rest.

JCW: In the forestry and the biometric literature, there are enough flexible forms to the equations. People all over the world have had to deal with trying to predict things a lot of different ways, and so there's some good flexible forms of equations. It was just fitting the coefficients to our particular situation.

HKS: As an economist, you think about efficiency in certain ways. Do economic theorems have to be modified when you're dealing with Third World situations or socialistic situations?

JCW: What comes to your mind?

HKS: Our idea of efficiency—the hidden costs of labor and all that. Do the First World models really work in the Third World by just putting different numbers in? Or do you need a whole different set of equations? I've talked to some people who in the '60s were in Africa working for the State Department, on foreign aid. A great mistake we made as a nation was to apply the theorems of the Marshall Plan, which was very successful in a demolished First World part of Europe. The same philosophy didn't work well in Africa, because it didn't have a First World mentality. How else do you have to think differently when you are in the Third World?

JCW: Economics is a big profession. When people think of economists, they tend to think of the mathematical economists. Those are the folks that get the biggest press and get in the journals and so forth, a lot of times. But I don't think we can forget that economics basically is a social science.

I'm saying all this to get at your question, that in any particular institutional setting the mathematics may be the same, just like I was talking about flexible forms of equations and biometrics. How you apply those mathematics and what the appropriate mathematics to apply are, can be quite different because of the institutional setting. In some settings the mathematics is not the appropriate tool even to analyze your situation, your economic situation that you find yourself in. What you're describing with the Marshall plan, I guess, has to do with the incentives, you know, what are people trying to maximize in their individual utility functions? Probably there were some assumptions made that the utility functions of individuals in Africa would be the same in Europe. They're not, because they have different assets to work with, they have different social heritages, and a lot of other things.

I would say in Brazil, getting specific to Jari, with respect to analyzing plantation investment, probably net present value analysis and all that sort of thing was very similar. Tariffs and taxes, that was peculiar to Brazil, and you had to work with that, which of course is part of the price mechanism, how you interpret that. I think there's a lot of the economics of working under uncertainty that we had to deal with, too, risk and uncertainty. You think about Mr. Ludwig, one of the classic assumptions in economics is it assumes profit maximization. Certainly Mr. Ludwig would say, I'm trying to maximize my profits. I mean, that's why he got into this. He wouldn't have said, Well, I'm getting into this to lose money.

HKS: O.K.

JCW: But on the other hand, I think he had some other motives. Obviously we can't talk to Mr. Ludwig to find out, but I think he wanted to leave a legacy beyond just generating a lot of cash flow. I can't help but believe that, because I think there were other places that he could have made a lot more money, and I think he's a rational person. He could have seen that. There was something that really struck him about this project, that was his love, and he was willing to continue to put money on this project in the face of some pretty big obstacles. So I think profit maximization is not the correct model to apply to the Jari project and his motives.

HKS: I suppose a good economist, like a good anything else, any other profession, is realistic and so forth. What I was thinking about when I was asking those questions is, looking at the United States, there's a group of economists who insist via the Sagebrush Rebellion sort of thing that we'd be better off if public lands were privately held, that under private ownership there's better stewardship because the market forces tend to make you want to protect your investment.

JCW: Yes. Somebody owns the resource.

HKS: That's right. If you go to Brazil with that idea in your mind, are you handicapped?

JCW: I don't know about the Sagebrush Rebellion, but I basically have believed that the free market system is not perfect, but it is better for somebody to own the resource, and they will do a better job of managing it than if everybody owns the resource, sort of the tragedy of the commons argument.

HKS: O.K.

JCW: If nobody owns the resource, the resource will be abused. There's a lot of mathematical theory to show why that is. If no one owns the resource, the resource will be depleted, whereas resources owned by one individual or one group of individuals sharing the proceeds from that resource, that there'll be a better solution for the economy as a whole. Whether it be Mr. Ludwig owning that resource or whether it be a group of companies like it is now, that's a better solution than Brazil owning that resource and managing it as a socialized enterprise.

Government's role is to put the appropriate regulations and taxes and tariffs, to take care of externalities. This is what I was talking about to get a good agreement with the government, the government doesn't need to be changing the rules in the middle of the game. They need to make the entrepreneurs understand what those rules are so that the entrepreneurs can react appropriately. If the government keeps changing the rules of the game as you go along, then you're going to lead to suboptimal behavior. The people are going to react in bad ways. The private sector's going to react in ways that neither the government nor the private sector is going to be happy with. So I guess that's why I'm saying that there needs to be a clear understanding what the rules of the game are between the private sector and the government, when you enter into something like this. But we know that doesn't always happen.

HKS: I realize your training in economics was less sophisticated when you were there than it is now. And you've had more time to think about it and all the rest.

JCW: Yes.

HKS: But did you ever engage in these sort of economic theories, I'll call them, with a Brazilian colleague, who comes out of a different system with different expectations.

JCW: This is like me saying what the average American thinks.

HKS: Sure it is.

JCW: I would say there's not a lot of difference between the average American middle class, upper middle class, who after all are the people that are really managing the economy. To a large degree, the entrepreneurs and the managers in the economy are the middle class or upper. If you take the average upper middle class American and the average upper middle class Brazilian, I don't think you could tell the difference, in terms of their attitudes toward government and the economy and what the role of government is. I often think there's a lot of parallels between Brazil and the United States in the people. We're just a little different. Obviously there are more people who are economically disadvantaged in Brazil than there are in the United States at this point in history, but we're both countries blessed with a very big resource base, forests, minerals, all that. We're both countries with a very heterogeneous population in terms of ethnic origin. At the same time we had an immigration from Europe to the United States, they were having a parallel immigration to the southern hemisphere, and that's where a lot of the immigrants in towns like Blumenau in southern Brazil, primarily German, came at the same time as my German ancestors came to Pennsylvania. There's a lot of similarities to me. I don't know that you can see there's a lot of difference in economics, philosophy.

HKS: I'm trying to develop a basis for people who will use this transcript at some time in the future, in judging the success or failure or the goodness or badness of Jari. Obviously Jari is in Brazil. It belongs to Brazilians in the long term, it's part of their culture, but how does one measure? There's so much criticism of it. How are we going to sort this out?

JCW: It gets to the issue of multinational corporations or whatever. If Mr. Ludwig buys Jari, supposedly, it really belongs to Mr. Ludwig. Now granted, he has to work within the environment of Brazilian laws, and I think that's part of the problem. We had the *abertura*, the democratization. Where Mr. Ludwig and his folks were blindsided was the fact that the Amazon holds a special place in the hearts and minds of Brazilians. Even though Mr. Ludwig can say, Well, I have the title to this, or I have as good a title as you can get in Brazil with their various land problems with title and so forth, you're still going to leave yourself open to political pressures, outside of the economics, that say, Look, you may own that, but this is Brazilian. I think that's a problem that faces anybody, whether it be Japanese coming to Tennessee, or whether it be Americans going to Brazil, they've got to deal with that issue in some way, either by good public relations or with other things. If one supports transnational and global economics, it's a challenge for that way of doing business and the transfer of capital between things, because unless the entrepreneur can have the ownership of the resource—they don't want to be subject to the political problems, so to speak.

I'm thinking of the United States as much as I'm thinking of anyplace—one controls an asset in two different ways. One controls an asset by agreement within the community where one finds itself. In other words, there's sort of a common knowledge, this is your asset. But the other way one controls the asset, and this can go for your car or anything else, is legally. This is my car, or this is my piece of land, because I have a title in the courthouse.

The trouble in Brazil and even some places in the United States is that if the common knowledge that this is your asset breaks down either because politicians or the local community says this is no longer yours as far as we're concerned, then you have to go to the legal side of things to defend yourself, where the lawyers get involved and all that sort of thing. That's I think what happened with Mr. Ludwig. He had the same difficulties that a lot of other people have in Brazil with legal title, and because of the pressures, in part the pressures because of what I would consider a lot of demagoguery and the populist movement, he had to try to resort to some extent to the legal title, legal things. There's some real problems in Brazil as there are in other places in the world about legal titles. And the lawyers can kind of do what they want to with that. You just go before the judge. I don't think it ever got to that case in Brazil. That may have been one of the reasons that Mr. Ludwig decided to leave. But I think there were other reasons, for example, not being able to get the government to give him approval to some things about infrastructure. There were some other things that caused him to decide to pull up his stakes and to sell out.

HKS: When you were socializing with your colleagues there, or you had the consultants, say the people from Weyerhaeuser came down, and you were working with them, did you ever get into any sort of theoretical discussions as a part of that? Or was it purely technical, how to get the job done?

JCW: I didn't talk to them. My job there was to get the job done. I could have my own thoughts about this whole thing. When I faced myself every night, you know, or looked in the mirror and said, Can I live with myself? Am I doing the right thing, my feeling was I was doing the right thing because we were giving people jobs and we were developing that economy there. I really didn't discuss the philosophical issues, whether this was the right thing in the great global context of Brazil or Brazilian politics.

I think the biggest conversation I ever had about something like that was with Michael Stanton, he's at a university in Rio. It had to do with Brazilian policy about letting PCs come in. We were discussing—he is English, by the way—whether it was a good policy for Brazil to keep out imports, because the positive side of keeping out imports, the infant industries type argument, was that they could allow their infant industry to develop. I would argue on the other side, though, that as a person who needed those computers to do his job, that there was a real cost to be paid, because people were not being able to be as efficient in their individual industries, forestry or automotive, or whatever, because they couldn't bring this new technology in. It was being blocked. You were being either forced to use the old technology or inferior technology.

Once we'd moved out of Oklahoma State, where we did our data processing, we found a place in Rio. He was very good at working with us and he did some of our data processing at the mainframe computer in Rio. We got into that discussion because that affected what I was trying to get done directly. That was the government policy that was bothering me, because I felt there were better ways to do this. I couldn't, of course, control any of it. But it was an interesting philosophical argument.

HKS: This link to Oklahoma State, does that go back to Posey coming out of Oklahoma State? Did he happen to know somebody there?

JCW: First of all we looked at Belém. There was a computing facility in Belém with an old IBM. Maybe it was an IBM service center in Belém. We thought of doing our computing there, but the computer was just so old and antiquated. Our next thing was to say, Well, maybe we need to find a university facility in the United States that we could use their mainframe. I looked at the University of Georgia, VPI (Virginia Polytechnic Institute), Oklahoma State, and the University of Florida. I really didn't see any significant differences between any of those, and Clayton had ties to Oklahoma State.

We had used a consultant from Oklahoma State who'd been a professor of Clayton's and later became a good friend of mine, Professor Nat Walker. We decided to use Oklahoma State because of some of those connections. At one time we had two people working at Oklahoma State that worked for me, J. L. Albert, whom I mentioned, and Sam, I can't remember his last name. It begins with a W, but anyway, I had two people working there at Oklahoma State, doing some of the biometric data processing work. Then eventually I wanted to try to move that closer to Brazil, and particularly as the project started trying to downsize a little bit with the possible changeover in ownership. So I got in contact with Michael Stanton in Rio.

HKS: Today, with the communications, you could downlink, uplink anywhere in the world. It doesn't much matter where the facility is. It mattered more then, I suppose.

JCW: I was interested in the data processing from a forestry standpoint. The company had a data processing problem with respect to all their other records, accounting records and all that. There was a fellow named Charlie Schick who worked for the company, and I think he was based out of Rio. The company was actually looking at satellite links and telephone lines, how can we use mainframes? Nowadays, this is sort of a nonissue. That's a real advantage that anybody starting a new project in a developing country, they can just go straight for PCs to do everything they do in forestry.

## Forest Operations Planning

HKS: Forest Operations, Planning. Your job was to coordinate forest management and logging plans with bleachcraft, market pulpwood. Coordinate with who, is the question I'm asking. This is within the company. Right?

JCW: Usually with Johan and Mac Davis or John Sessions from Harvesting, we would attend a weekly meeting down at the pulpmill, where Ted McCrocklin was the pulpmill manager. We talked about wood deliveries for the next week, and plans for mill shut downs, and all that sort of thing. So we would coordinate with the pulpmill, their production. Which is a very common thing in pulpmills, to have a weekly production meeting, and usually the people in woodlands are part of those meetings.

HKS: I don't know very much about pulp, but it strikes me as significant that you had three basic species. There would be a desired recipe, I'm assuming, a percentage of eucalyptus, pine, and so forth.

JCW: Right.

HKS: Was that a difficult thing to maintain, the flow, species by species?

JCW: Start with the marketing of the pulp itself. This was basically market pulp. Everything started with the marketing and sales people. They would develop a customer base, and the customers would want a certain type of pulp. I think we had two or three grades of market pulp. We had what was called the pine pulp, we had the gmelina pulp, and then we had what they called Jarilyptus, which was a kind of a eucalyptus pulp. That was mainly in the planning phase. So it started with the marketing, and then the pulpmill people had to schedule their production of what to produce when. Then we in the forestry had to figure out how to deliver the appropriate amount of wood, during the year, of those species.

One of the issues that concerned us, in coordinating the delivery of that wood, was the wet season versus the dry season. Typically gmelina, which was the premium one of our pulps, it was better if it could be produced in the dry season, because of the type of soils. Some of the soils were very difficult to harvest in the wet season. Also, if you got out in the wet season, you could cause harvesting damage to those soils. So if you were looking at it from a woodlands perspective, the ideal thing would be to produce pine pulp in the wet season, and produce the gmelina in the dry season. But that was not, of course, what your customers necessarily wanted. There were the coordination issues in determining what harvesting sites to pick, and how much wood inventory you needed. On top of that, we were using a certain percentage of native species in some of our pulp. In coordinating the native harvesting and doing the selection of the different native species—I think there were about sixty different native species we had determined could be used.

Another part of the weekly process was working with the lab technician at the pulpmill, Beatrice Redko, who was doing some of the testing to see if new native species could be added without harming pulp quality. There were oftentimes disagreements between Ms. Redko and the mill manager as to whether it was an appropriate time to try to move in a new species, the risk involved in quality and all that. And as in many pulpmills, there are oftentimes disagreements as to what, if you have a pulp quality problem, what is the source of that problem. Oftentimes they looked for the wood to be the problem, but other times it could be that there's a problem in the mill itself causing contamination. So those were the kind of things that got discussed at the weekly meeting.

HKS: I'm assuming that in the world market that this stuff was sold into, there was another layer of customers, and they had a preference. Did the mill management get a message from the paper industry in Sweden, they want so many tons of gmelina pulp delivered by a certain date? Then that would go out to you?

JCW: We were completely out of that loop. Our customer was the pulpmill. Now the pulpmill would find out from marketing, to do their production planning. If marketing said, Well, we've got orders lined up for gmelina pulp, then the pulpmill has to figure out how they're going to produce that gmelina pulp. One of the ingredients to that production is to have the gmelina there to produce it. Then we would try to work through how are we going to get that gmelina to them, and send them the appropriate signals to say, Well, we're going to try to get it. But if you can send signals up the line and say, It's difficult for us to get gmelina this time of the year. So can you and marketing and sales, maybe, work with the customers and reschedule things? Or maybe we need to have an inventory of pulp someplace.

Market pulp is a very cyclical type business, where prices go up and down. It's a low value kind of product, as pulp derived products go, and a commodity. So that adds on to the scheduling problems of the marketing and sales side.

HKS: It must be significant as time went on, and you phased out gmelina, in terms of responding to the world market.

JCW: We didn't phase out gmelina while I was there. We were still producing a lot of gmelina.

HKS: I'll put it a different way. You didn't feel under some kind of pressure, or you weren't disgusted after one of these weekly meetings—those people in marketing have no sense of where pulp comes from. It comes from the ground. They were selling stuff that you couldn't produce. You didn't feel pressures like that?

JCW: No, not in the short-term period. It was the day to day timing issues, or week to week timing issues. I do think one of the significant sort of strategic things that we faced, though, was the limitations of gmelina as a species, given our soils. Here we had a species that we had really marketed. That was our marketing thing, gmelina is our pulp. It had a lot of good properties and so forth. To have to go to the marketing people and say, Look, we need to think about this, because we can only produce a certain amount of gmelina. Or if we produce more gmelina, our productivity in the forest is going to suffer unless somehow we can come up with some better genotypes and some other management strategies. So there were some strategic marketing issues with respect to being able to grow the kind of trees that they wanted, that were discussed, and gradually led to the demise of gmelina, because it was really driven by the forest and what we can produce, as to what we can produce in the forest and what we can sell in the markets.

HKS: Gmelina was the pulp that was desirable because it's clean?

JCW: I'm not the best person to ask, but it's a very white pulp, short fiber, like cottonwood. One of the favorable characteristics of it was that it had a fiber that, for its shortness, was fairly strong. Most short-fibered pulps are not very strong. The fibers collapsed in the processing sort of like you'd think of a straw. If you have a bunch of straws together and the bonding is just between the straws and the straws were still circular, they're not going to have as much surface area to bond. But if those straws were to collapse, there's going to be more surface area to bond. I think gmelina had some kind of a property like that. All of this can be documented based on pulp trials, and there's a lot of pulping that's been done with gmelina, as I say. Other people are now using gmelina.

HKS: The technical question could be answered by going to any good library.

JCW: Or talking to somebody who's currently processing gmelina.

HKS: But your marketing people would have preferred that gmelina had been a good species?

JCW: Right. Eucalyptus is a good pulp, too, and that's what they're growing now, but gmelina had some characteristics that were in some ways favorable to eucalyptus. Now probably in some other ways it was unfavorable to eucalyptus. We were hoping it would have its own niche in the world markets. But then you ask yourself the question, If you can't produce a lot of it, do you want to spend a lot of energy trying to develop a niche for a very limited amount of supply that you can produce?

I think the company after Mr. Ludwig, and for the year I was there, and then for a few years after that, they had not given up on gmelina. They felt that with more intensive management of gmelina, they could really produce a lot more per hectare, even on some poor soils. And who knows? Maybe if they had kept at it with better genotypes.... I wasn't there to observe the reasons, they've now adopted a strategy of eucalyptus. One of the advantages of eucalyptus is it's not as site sensitive. There's a lot more genotypic variation that can be taken advantage of for different sites. There's a lot of knowledge about eucalyptus in Brazil, and the management of it, and what works and what doesn't work. Whereas with gmelina, you're sitting there with a species that you're one of the few persons that's working with it. So for whatever reason, the strategy right now appears to be to go with the eucalyptus as the market pulp of choice.

HKS: To carry that a step further, to work with native species would be even more of a gamble? I mean, gmelina was a tough enough thing, but no one really knew anything about the local species.

JCW: Yes. But the advantage, you had a relatively cheap source of wood. The cost was just the cost of harvesting it. You had three uses of the native species. You had fuel, you had lumber, and you had pulping. Not every native species could be used for all three of those things. In fact, most couldn't. Some could only be used for fuel, there weren't any markets for lumber. Everything that could be used for lumber or pulp could be used for fuel. There were some of the fuel species that could be used for lumber, and some of the lumber species and some of the fuel species that could be used for pulp. So the trick was to try to find the highest and best use for those different species. You had a cheap source of fiber, and you wouldn't develop a pulp for this native species, but you would blend that in with the other without sacrificing the quality of your Jari pulp, your gmelina pulp.

HKS: Were you able to do this testing at Jari after the mill was put in?

JCW: Yes, after the mill was put in. We probably have some of the best data on these different native species in terms of their pulping qualities. I haven't kept up with that. I think there's some other projects that were looking at that, too. But of course, now I think rightfully so, concern about using the native forest, what the role of the native forest is for biodiversity versus wood production, maybe this is all a moot point. Although we did also have native species trials where we were growing native species, too. When I visited Jari recently, I saw some Brazil nuts that we had planted back in 1970 down in Munguba near where the pines were planted. Those are now thirty-six years of age and about two feet in diameter. And there are places in Brazil where they have plantations of Brazil nut trees. I think you can produce Brazil nuts in about eight years. We interplanted Brazil nuts in some gmelina plantations. I don't know what's become of that, where we planted the gmelina and then we planted Brazil nuts at a wider spacing, with the idea of perhaps having Brazil nuts above and the gmelina plantation below. What I'm trying to say is we did have planting trials of native species that were established. Currently they also continue some of those trials.

HKS: Most of this is hearsay on my part, but Johan is managing in Belém the Tropical Forest Foundation project.

JCW: Yes, I know that.

HKS: By my understanding, I would call it low impact silviculture as opposed to plantation silviculture.

JCW: Yes.

HKS: Do you have any sense of does that work? Can you supply pulpwood to a mill like Jari with low-impact silviculture?

JCW: I think it has a niche. I think that it could be done. I've seen one of the videotapes that Johan has done with the Tropical Forest Foundation. He's testing the operational feasibility of doing that on a small scale. I think the next step, at a place like Jari, or elsewhere, would be to say, how would we fit this into our operation plans, given our other sources of wood? I can easily see a situation where you would have plantations as part of your wood mix, but you would also have native forest. Now there are some problems. I mean, you want to maintain diversity of all these different species, and you've got three hundred and fifty different species. How do you bring these things in and do the merchandising and all that? It really comes down to wood costs. Can you deliver wood fiber from that silvicultural method, at a competitive price to the mill? It depends on a lot of things.

HKS: The price of market pulp is significant in land management. That's really the end product.

JCW: Yes. The model I would think of is you would probably want to look at some lumber. You'd want it to go to the highest value use. If you found these as lumber, you want to bring them in as lumber, if that was the highest value use. As an interesting aside about the highest value use for the native species, I've heard Jari criticized in the popular press, Well, it's a shame to burn up these beautiful tropical hardwoods that could be used for lumber.

HKS: Sure.

JCW: There are a couple of problems with that. Many of these species can't be used for lumber because they're so dense. The other reason they can't be used is they look beautiful on the outside, but they've got a lot of rot and decay on the inside. But another reason is because of the high cost of fuel in Brazil; we did some calculations one time that showed the highest valued use in terms of economic use was for fuelwood, not for lumber. Which is a surprise to a lot of people. We're somewhat shielded in the United States from what a lot of other countries encounter in terms of energy costs.

HKS: Right.

JCW: That's an interesting spin; wood as a fuel is a pretty valuable commodity. Now, the other criticism, and I think a legitimate one for discussion, is maybe the highest and best use for those trees is standing, because of their contribution to biodiversity and so forth. But getting back to what Johan's working on, my own personal opinion is that he's doing a good service by trying to look at some sort of intermediate, where we can not only get some of the aesthetic and biodiversity benefits, but also try to tap into some of that resource for some cash returns for that economy, from foreign economy.

HKS: At some point, this becomes philosophical. What right do we have to do any of this, anyway. That's what you read in, I'll call it the environmental literature, without defining what that is.

JCW: That's right.

HKS: The Amazon should be a preserve, like Alaska should be a preserve, or whatever.

JCW: Yes. We know that it becomes real philosophical about what right do we even have agriculture in the southeastern U.S.? Maybe we ought to just all go back to Europe. You can get all kinds of philosophies.

## Silviculture: Jari and Southern Brazil

HKS: One of the things you wanted to talk about earlier, and we only touched on it a little bit, was comparison of silviculture in Jari to silviculture in southern Brazil. I'm not sure what the significance is, but let's talk about that as much as you want.

JCW: The period of Jari I'm referring to is the period from say '72 to '80, and to some extent even today. At Jari we took an extensive approach to management in silviculture, very similar to the United States in the sense that you would go out and you would clear an area, you would plant it, you would weed it to the extent that you needed to liberate the trees and grow them. In southern Brazil they took a more mechanized approach, a more agricultural approach to it. In part that was because of the nature of the land base they had to work with.

We were working with a land base which was primarily native forest. Some of the mechanization that was done early on up at the airport caused soil damage. I think the cheapest and the best way was to burn the area, leave the native stumps in place, hand plant it, hand clean it, and so forth. Now, when it got to the second rotation in Jari, on some areas, we would start windrowing again to try to clean up—we started getting ready for a mechanization phase, where we could mechanize. And as time has gone on, it's evolved to where things can become more mechanized, as in southern Brazil.

In southern Brazil, a lot of those plantations were established under an incentive program. Brazilian tax dollars paid for the establishment of those plantations. Until the very end, we didn't really take advantage of any of that. Or maybe we couldn't. Anyway, we didn't get those kind of subsidies that they did in southern Brazil, plus the land base they were working with many times was degraded agricultural land. It was second growth forest, or second growth bush. What you see about plantation forestry in southern Brazil started off with where Jari is sort of today, with the ability to do mechanization.

We were down the evolutionary ladder a couple of steps in terms of silviculture. Sometimes when we had visitors from either southern Brazil or from the United States, who would look at our situation, I think we were criticized. The benchmark was southern Brazil, neglecting to look at what was appropriate for our situation was a little different from what was, what the land situation, and the incentives, and the money being given in southern Brazil were.

HKS: As the enterprise shifted from being an American operation to Brazilian operation, the Brazilian experience in southern Brazil would tend to be applied because that's where the internal expertise was coming from. It would have more influence under Brazilian control than it did under the U.S. control.

JCW: I think that's in part correct. I think also it was partly because of the evolution that I described. If you had spent your money with the mechanization at the front end from that native forest, pulling up those native stumps, you would have spent a lot of money. Whether you were Brazilian or American or wherever your technology was coming from, I think it would have become very apparent that the net present value would be a lot lower by having to incur that up-front cost of windrowing and everything else.

The other issue is, when you're doing that kind of windrowing, you're doing a lot of soil disturbance and compaction. In order to move these native tree stumps and so forth, you're going to be compacting. Those tractors are going to be on there a lot bigger period of time. So I guess what I'm trying to say is, as there was more Brazilian silvicultural influence in this—and the Brazilian plantations in the South, in technology, were evolving during this period of time, too—it would be natural for more of that to take place.

But I think the other thing that happened was that we were ready to do it because we were in second or third rotations. If you were to talk with Robin Collins, he could give it a very unique perspective, because here was a person, Robin, who had worked with Westvaco in their plantations in southern Brazil. He came to work for us with that background. He came out of a whole different setting, so I remember us having some discussions internally, in forestry operations. Should we mechanize more than what we're doing? Beginning in about 1979, we started doing some windrowing again on some of the second rotation stuff, trying to prepare ourselves for more mechanization and less labor intensive approaches to the problem. We never did that even in subsequent years to the first rotation in plantations, because of the high cost. High cost not only monetarily but the high cost in terms of topsoil damage and that sort of thing.

HKS: You've said, in one of your papers, that broadcast burning is probably the single most effective technique for site preparation in the humid tropics. Given the tremendous negative publicity, the satellite imagery, everything picks up all these fires in Brazil—were you concerned with that?

JCW: Concerned with what, the publicity?

HKS: The publicity.

JCW: No.

HKS: It was the best way to do it, and that's what you did. The general perception is that this is a bad thing for the Amazon, all these fires. This is the way it's portrayed.

JCW: Let's dissect that. Why do you think that people think that's a bad thing?

HKS: I'm not sure why. Maybe it's just symbolic of the development of the Amazon, and anything that develops the Amazon is bad.

JCW: I think that may be the case. I guess what I'm saying, if you're against cutting down trees or burning trees, anything you do, whether to establish plantations or grasslands, or whatever, is bad. I'm not talking to you, but you could just end the discussion there. Now given that you are going to establish plantations, or grassland, or whatever, I stand by the statement, the most effective site preparation method in that situation is burning. One way to look at that empirically, if you look at slash and burn agriculture, that's what the local folks use.

HKS: Sure.

JCW: Even if they can go to somebody who has a tractor, they don't say, Well, hey, come on over and plow my land up here with the native forest, where there's slash and burn. I know that they don't have access to that, but I would say even if they did have access to it, they're going to determine that's not the cheapest way to do it, it's not the best for the site, for the subsequent crop, because they're going to go in and mess up the topsoil, you're going to compact the soil. I stand by the statement that the single most effective way to do it is to burn it. You've got all that ash layer, which is fairly basic, and you're dealing with acid soils to begin with, so it provides some buffering capacity. Whether it be trees or manioc, it's probably the most effective way to do that job.

HKS: I'm thinking of the mess I've got in my yard in Durham because of the hurricane damage. It's going to take a year before that stuff is dry enough to burn. Did you knock it down one year and burn it the next, or what?

JCW: If you didn't let it sit, if there was no more than about a two month gap between when you did it, or a three month gap between when you did it and you burned it, there wasn't any problem, even though some stuff had greened up. There were two issues. One would be, you would want the wood to be dry enough to burn, the vegetation, on the one hand. The other side of it was that you didn't want too much vegetation to regrow, to sprout back, because contrary to some popular literature, and people that have field experience, no matter what their viewpoint about burning or not burning is, know that when you cut in the tropical rainforest, it comes back like gangbusters. Now it's not maybe the same stuff that was there before, but there's a lot of vegetation that comes back, unless you do something. So if you did it within two or three months of clearing, you had no trouble burning. There was enough biomass there.

HKS: It was dry enough.

JCW: It was dry enough.

HKS: You could ignite it with a reasonable amount of diesel oil or whatever you put on it.

JCW: The method we used was, for the most part, sticks that were about six feet long and they had burlap of about one foot long wrapped around the ends of them, and they were dipped in kerosene. And you would start a fire by going around the area and torching it, and it would burn, so that the actual fuel oil was just to get it started. Which is no different from controlled burning in the United States or many other places. You didn't have any trouble burning it.

## Return to Jari

HKS: You've been back to Jari in April of '87. You were a consultant?

JCW: Yes.

HKS: They hired you to come down and do something?

JCW: Yes. My report was about the growth and yield system and to give my evaluation of some silvicultural things and like that.

HKS: You'd been gone five years.

JCW: Yes.

HKS: Were you generally pleased, or satisfied, that the work you had done was being continued?

JCW: Yes. I think they were progressing. Consultants. I always felt when consultants would come down, when I was working there, you were paying them to tell you what you already knew, or you were paying them to be your spokesperson because a prophet in his own land, you know, can't work. So you had to have them say what you already knew. Or you were giving them all the information so they could write reports and make money, O.K.? So I didn't really want to be that kind of a consultant. All I did was to go down and give an independent opinion of what I saw, given my perspective of being there, and I felt like they were going down the right path.

I was a little concerned that from a species standpoint they were deciding not to develop pine very much any more. They were going straight to either gmelina or eucalyptus. I believe that pine has a place in plantation forestry at Jari. I think it grows well in poor sites. I think the lacking thing is that more genetic work has to be done with it. The company had a seed orchard, still does, I guess, at a place called Morado Novo in Minas Gerais, with pine jointly set up with Weyerhaeuser, and were going for first generation improved seed, based on genotypes or phenotypes that have been selected there at Jari. If you go to those plantations, you can see a tremendous amount of genetic variation. So I just feel like pine is the species that needs to be looked at. They were I think mainly for cash flow reasons, maybe for some market reasons, going away from that.

HKS: Pine just doesn't grow fast enough? Is that the problem?

JCW: Well, first of all there are few tropical pines, that is, on the equator. *Pinus caribaea hondurensis* is not one of them. Its native range is in Central America. There's some photoperiod issues involved. You get foxtailing, growing on the equator. But there's enough variation. We had some provenances that did real well. Originally we thought the mean annual increments would peak at about sixteen or eighteen years of age, and we could grow solid wood. But what was happening is, it was peaking earlier than that, at twelve years of age. I'm sure it was genotypic. The other part may be nutritional, maybe based on some work we're doing in the Southeast with forest fertilization at mid-rotation, there may be some stuff that could be done to keep those things pumped up to grow longer. But I think most of it was genotypic. So what was happening, you were getting MAI peaking at twelve years of age, which some people might say, That's great! But on the other hand, it precluded you from growing solid wood products.

This is based on some work done at N.C. State with loblolly pine, which is not what we grew. You get tremendous growth rates of loblolly pine in southern Brazil, but that's partly because they have fairly cool nights compared to us up in the equator. So there's photoperiod, there's temperature, a lot of technical issues. But I think with proper genotypes, you could do a lot better at Jari, and there are places that are continuing to grow pine in the tropics. CAMCORE, which is an international tree improvement effort with tropical pines, based out of North Carolina State, continues the work for improvement of families and species that'll grow in the tropics. Back to my consulting, that was one of my criticisms, that I suggested they continue to look at pine.

HKS: But the rotation is twice as long as it is for eucalyptus.

JCW: Yes.

HKS: Economically it's a real question.

JCW: That's right. But on the other hand, we talked about in your tool kit, you should have a lot of different genotypes and species. If you were thinking about expanding to different products, if you could get better genotypes, you might go to solid wood products from pine. Oriented strand board, you know, there are all sorts of things. I just feel like it's a species that has a place. But you're right. That was the main reason, it's got a longer rotation. It required more silvicultural inputs. It was a very logical reason they went to eucalyptus, because it has a shorter rotation. There was other technology out there with eucalyptus.

HKS: Thank you for an interesting interview.

Tommy Thompson and 4-year old pine. Thompson
Photo.

# L. N. THOMPSON

## Introduction

Lawrence N. "Tommy" Thompson was born in Mt. Vernon, Georgia, on
October 31, 1925. He earned a bachelor of science degree in forestry from the
University of Georgia in 1948 and a master of forestry degree from Duke
University in 1950. Until 1973 he was employed by several firms —
principally Georgia-Pacific Corporation — whereby he became a hardwood
specialist. In 1973 Tommy and John Shackelford formed T & S Hardwoods
in Milledgeville, Georgia. Their initial investment of $200,000 has grown
into a solid company with annual sales in excess of $20,000,000 and a book
value of $15,000,000.

One of his earlier assignments for Georgia-Pacific was at its operation on the
lower Amazon. While there, he would hear about Daniel Ludwig's huge
forestry operation only a hundred and fifty miles away by air. Working
directly under his supervision was Johan Zweede; eventually Zweede would
be in charge of Ludwig's forest operations, and he frequently sought
Tommy's advice. By then, of course, Tommy was back in the States and

running his own company. But he made time to consult not only at Jari but also for Mead, Union Camp, and KMI Continental Group.

Tommy was a frequent visitor to Jari during its American phase and grew to admire and respect Clayton Posey, Bob Gilvary, and John Welker. He also greatly admired Daniel Ludwig, who too used Tommy as a consultant. As well, he admired the overall venture and became troubled that this great story was so poorly and inaccurately told. Thus, he spearheaded the effort to fund not only his interview but those of the three others as well.

# Learning the Hardwood Business

Harold K. "Pete" Steen (HKS): Let's start with your background so we will know what skills you had that made you attractive to the people at Jari.

Lawrence N. "Tommy" Thompson (LNT): I went to work for Georgia Pacific (G.P.) about two months after I completed my M.F. program at Duke, working on wood supply for the Savannah plywood plant. At that time we were still peeling a good many native hardwoods such as sap gum, tupelo, yellow poplar, and so on, making pretty much conventional hardwood plywood. My work was primarily involved in cruising hardwood timber. Also purchasing logs from producers that might control their own stumpage. At that time a lot of the logs going into the Savannah plant were shipped by rail. Most of them in gondola cars.

HKS: I've cruised softwoods in the Pacific Northwest, never hardwoods. Is there a big difference?

LNT: Oh, not really. I don't think a big difference.

HKS: You have to know more about grades?

LNT: Yes, you probably had to be somewhat more concerned about grades and defects. Particularly in the rotary veneer business. Any time you've got knots or hollow in a tree it makes it impractical to peel. I used to carry a small hatchet with me most of the time, a single-bladed hatchet. Bang on 'em and really got where I was pretty good at detecting if it was hollow there. It was a good, broad experience in forestry procurement. Not for really forest management although occasionally a tract would come along where I would participate in the way in which it was cut and left and so on.

HKS: But for hardwood management, it's sprouts. You don't plant. You don't get seedlings and plant them.

LNT: No, it's pretty much economically impractical planting hardwood and fortunately, it's seldom needed.

HKS: Do people grow hardwood on purpose?

LNT: Yes, and that's coming on more and more. Southern hardwood forestry has not had the attention from the forestry schools I think that it could have had over the past thirty years. Most of your big industrial forestry land owners, Union Camp, Georgia Pacific, International Paper, many of those big companies have been primarily concerned with softwood, specifically southern yellow pine. Mississippi State has come on pretty strongly in really the last fifteen or twenty years with greater emphasis on hardwoods. It's a more complex silviculture in that you're rarely going in for a complete clearcut.

HKS: If you go south from here a hundred miles, the hardwood is not a furniture grade. There's a change as you get into the lower elevation. I'm not sure what it is.

LNT: It's a question of drainage, soil types, and so on. There's some down there in some species that's still of furniture quality. Yellow poplar will turn out just as well.

HKS: Is this something you can see from the outside, or you just know it through experience that when you get a certain place it's no longer high quality? What I'm leading up to is, when you go down to the tropics and you start looking at the trees, what do you see?

LNT: There are most always indicators that give you a clue on the overall quality and health. I would say one of the easiest ways to get into difficulty in buying hardwood for commercial lumber production is getting into species that are overmature and the indicators normally being the crowns. When you see a good full crown, no dead limbs, no streaks of dead bark on the bowl or anything, you're looking at the best.

HKS: Do those indicators work in the tropics? The same basic principles or is it a whole new ball game for you?

LNT: I was continually asked by people who saw my slides showing some of the indications of very heavy forest, why there wasn't more done at Jari to salvage more of the timber rather than burning it up. One of the big problems was that much of that three hundred thousand acres that was cleared was overmature and had a good bit of defect in it.

HKS: We'll come back to that. I'm Ludwig and I want to hire an expert at a variety of things. I hire an engineer out of Cornell [Robert Gilvary] who's done some highway work and send him to Jari.

LNT: Yes.

HKS: Ludwig picks these people, so your background here is significant.

LNT: Initially, Johan Zweede and Clayton asked me to come to Jari. I got acquainted later with Mr. Ludwig and actually did some work directly for him.

HKS: How did you know Clayton and Johan?

LNT: Johan worked under my overall direction for four years at G.P.'s operation in Portel, Brazil, which is about one hundred and fifty miles by air from Jari. I had not met Clayton until I made my first trip to Jari, but I knew Johan very well.

HKS: So it's no surprise that you were invited, because you were already working in Brazil and you had hardwood experience.

LNT: Yes. Do you want to get a little more into my background?

HKS: I do. One of the things we need to talk about is hardwood lumber. I think that's probably significant because of the decisions that were made in Jari about not using native species.

LNT: Right. I spent twenty-one years with Georgia Pacific, and throughout that time I was involved in forestry procurement, particularly buying a lot of timberland. In the span from 1959 to 1969 we took our land ownership in the Southeast, and that included West Virginia as well, from less than ten thousand acres to almost nine hundred thousand acres in ten years. I spent a lot of time working in that area. During roughly the same time span I had charge of all the hardwood lumber production G.P. had which was at that time all here in the Southeast. We had some operations west of Gainesville that were not doing well. The CEO, Bob Pamplin, asked me if I would take on a project to turn those operations around. When I went down they were losing roughly a million eight hundred thousand a year, and we turned it around to about three to three and a half million dollars a year profit.

HKS: What was the deficiency that you had to correct? What wasn't being done?

LNT: Primarily management. Morale wasn't the best. Some technical things, but I'd say primarily my major contribution down there was to really give the employees and our local communities around us down there credibility for Georgia Pacific. There had been a number of things allowed to creep in down there that shouldn't have.

HKS: As a stereotype, a major corporation has various divisions that has people that look at the quarterly reports coming in from the divisions, and they quickly spot something that's not working. Apparently they didn't know how to fix that for quite awhile.

LNT: The work and decision making process in Florida was split up, and this manager handled the plywood plant and that manager handled the forestry. I was in charge of forestry all along. I was given responsibility for the whole thing except sales and was able to pull the people together or in many cases get other people in.

## With Georgia Pacific in Brazil

HKS: Management obviously was an issue with Jari. Lack of continuity at the very least.

LNT: It was amazing. I knew about Jari almost from its inception. In fact, the block of property at Jari was offered to G.P. I was not supposed to have to get involved with Portel or in the Amazon, but later I had to. Pamplin and Hiram Mersereau, Mersereau being my boss and Pamplin his boss, concluded that really for what they wanted to do in the Amazon, Jari didn't fit. They were primarily interested in the species called virola. It was used virtually one hundred percent for core stock for high grade plywood with birch, walnut, all of that.

HKS: So it was peeled or sliced there and shipped to the States for manufacture?

LNT: It was all peeled, rotary cut, shipped directly from the plant site. We had deep water there, into Port Wentworth, Georgia. By then the same plywood plant which I had started out for in 1950 was no longer peeling any domestic veneer.

HKS: What makes good core stock? Dimensional stability?

LNT: Dimensional stability and peels smooth. Doesn't tend to make the panels warp, and virola is an excellent species for that. G.P. wisely did not buy the property down there at Jari. We bought some other properties with the veneer plant, at Portel.

HKS: So to operate in Brazil you basically have to buy property. You don't buy the timber from someone as you would in the States.

LNT: Yes. I was told early we were going to be looking in Brazil. G.P. said, we'd like to have you go, but I was really busy up here with several major timberland acquisitions plus operating responsibility. They said we're not going to ask you to go, but about six months later Mersereau walks in and says, "I know Pamplin and I told you we wouldn't need you down there, but we do. Get your passport." Anyway, my first trip down there things had been going about six months and they'd already purchased a pretty big block of land. Let's say six hundred thousand acres.

HKS: Pretty cheap land?

LNT: Cheap land. A dollar or two dollars an acre. But I started really digging into that. There was some serious misinformation that had been provided in the original study down there. I don't know what happened, but they projected stands of virola that would run eighty percent virola. In 1955 I had spent two months in the back country of what is now Zaire for Owen Cheatham, the founder of Georgia Pacific. In the real tropical jungles there, I had never seen anything like that kind of concentration of species.

HKS: This is a native stand of eighty percent?

LNT: Supposedly. That was one of my very first things I struck out to confirm or not to confirm and it just was not there. I didn't think it was likely to be there.

HKS: Can you manage virola by cutting and encourage it to come back and be dominant in a stand?

LNT: I guess you could in extreme circumstances. What we were doing there was primarily buying logs. Virola occurs more or less like tupelo or cypress here in the Southeast in low swampy areas. It was all hand felled with axes, hand bucked with axes, hand rolled to the water and made up in rafts.

HKS: So it would float obviously.

LNT: Most of it would float. Once in awhile you'd get a sinker. As long as they had water where they could float it.

HKS: G.P. owns this land, maybe owns it in quotes, but it owns the land and then that's it. There's no management plan.

LNT: There was no real management plan. Fortunately, it will regenerate pretty well.

HKS: So there's no Brazilian forest practice act, to use U.S. terminology.

LNT: At that time there was not. Now this was in the last half of the '60s. They have now put in requirements in their forestry operations that the industry has to utilize a broader range of species.

HKS: Oh, I see what you're saying. In other words, no more high-grading the stand.

LNT: Right.

HKS: Okay. What would a typical acre be? Ten, fifteen dominant species? Or a lot more?

LNT: A typical ten acres will often be what you just said, maybe even as little as six or eight dominant species, and then a half a mile away the ten acres may contain another six or eight different species. Very diverse forest, particularly when you begin to look at it on a fairly broad basis.

HKS: What's the diameter? These people are hand logging, rolling the stuff down to the river.

LNT: Most of the virola was not that large. A typical log that we got into the plant down there probably came out of a tree at eighteen inches to twenty-four inches DBH. It's hard to realize just how much water there is—creeks and so on—in that lower Amazon basin. When the river rises it spreads out over big areas. Typically up river at Manaus the water in the river rises, goes up to forty feet to fifty feet from the dry season to the wet season. But down in the area where we were operating near the Atlantic it rises typically five or six feet. A lot of times it would enable these fellows to get out there and darn near fell trees in the water and buck them up.

HKS: So they were skillful at using water. They'd wait until the right time of year.

LNT: Yes. Basically your travel is by water.

HKS: Was virola available at Jari? Is this a common species in the lower Amazon?

LNT: It's a common species in the floodplain and there's also an upland virola.

HKS: This floodplain. That's where Ludwig grew his rice and stuff too.

LNT: Yes.

HKS: The same kind of country then.

LNT: Yes. There's literally several million acres of that forest type from the ocean going up river. It occurs frequently for three or four hundred miles. Now I've not really had any firsthand experience further up the river than that, but most of the land at Jari was not swampland. It was rolling. It finally went up into the a escarpment where it was like let's say six or seven hundred feet high and then dropped off again beyond that. I had timber supply responsibility for Portel for four years.

HKS: How did Johan get selected, because he's the one who would bring you back in. How was he picked out of the crowd as it were?

LNT: Johan is a graduate forester from Syracuse. He was born in Indonesia of Dutch parents. His father had a major plantation on the island of Java, and they were overrun by the Japanese. Johan and his mother were put in one camp and his father in another work camp, and his father died in the work camp. Johan said when he was six years old he weighed less than he had when he was three. They almost starved to death and were finally liberated by Indian Gurkas. The plantation was gone. They came back to Holland and in fact I think even the queen put them up in one of her homes for awhile. I think later all of the family immigrated to the U.S. He had a good solid background in silviculture. He was completely fluent in Portuguese, and he was very good at operating with Brazilian contractors, labor contractors, clearing contractors, and so on. Clayton recruited him. To tell you the truth, they made a very good team. Clayton was the managing director in effect for about three years, and Johan was in charge of the forestry department. Of course, Clayton also had a strong background in genetics.

## Involvement at Jari

HKS: You're clearing up how you got involved with Jari.

LNT: Right.

HKS: It's very logical that you wound up there.

LNT: They primarily wanted me to look over their shoulder as a practical guide in logging, and they were already doing some lumber manufacturing there for use on the project. As you know, they built a tremendous number of homes and all of that. So I became involved with them and made my first trip to Jari in 1973. I had known about the project all the time.

HKS: You had a lot of hardwood experience in the States. Portel was a hundred miles away from Jari.

LNT: Roughly a hundred and fifty miles by air. It was south of the main Amazon and Jari was north of the river.

HKS: What were your expectations? You knew about Jari. You hadn't been there but you're certainly familiar with the basic lower Amazon ecosystem. When you got there were you impressed?

LNT: Yes. It was interesting. I'd hear about what was going on over at Jari. They were having a continual turnover of managers. They first started out mechanically clearing the forest. Somebody asked me what I thought. I said, it won't work. You'll destroy the soil fertility in those tropical soils, and it turned out precisely that way. Strangely enough, my son Larry went to Jari very early, when he was sixteen. Hiram Mersereau had a son the same age as Larry. G.P. had its own shipping, and we sent the two boys off on a trip down there with a ship with the explicit understanding that they'd go down and back by ship. Hi came in one day and says, "Tommy, I know we decided that the boys would go both ways by ship, but said, they'd really like to fly back." [laughter]

HKS: How long a boat ride was that? A week? In that neighborhood?

LNT: Maybe slightly longer. Larry said everything in the ocean passed them including sail boats. But they carried a big cargo of Caterpillar equipment down to Jari and off-loaded it. I think it was eighteen D-8s being shipped. A lot of other equipment. What did surprise me some—and it shouldn't have—was just how rapidly they were moving along with many things. Particularly including land clearing, which Johan set in motion, and the abandonment of mechanical clearing. Just a tremendous expansion of what's been done there for a couple of hundred years or probably longer—slash and burn.

HKS: Did they use chainsaws at Jari?

LNT: Yes.

HKS: At Portel did they chop, use axes?

LNT: Yes, but they still used a fair amount of hand tools on the small stuff, but the bigger trees they did go with chainsaws and so on.

HKS: But there was no windrowing. After the hand stage, they burned stuff in place more or less?

LNT: Right. What they would do was go through first and put the small stuff on the ground. You know, brush, small trees and all, and then come in and fell the big timber on top of that and burn it during the dry season, which is roughly six months starting September or October. You got enough mass there with particularly the small material at the base that you can kick the heat up enough that the big logs begin to burn and burn very hard. Bear in mind, maybe fifty or sixty percent of the material on the ground that's been felled is out of these very dense species like angelim. There's at least eight or ten species with a specific gravity of over 1.

HKS: So they won't float?

LNT: No, they wouldn't float. We used to bring massaranduba timbers into Portel. For storage we had clear water at Portel, and we just dropped them over in the bay where it wasn't too deep and left them in there until we needed them. They'd sink just like a rock, but that also makes that kind of wood pretty easy to burn. Once you kick it off it'll go right on.

HKS: And if you don't burn it, it's there for twenty or thirty years probably. It doesn't rot.

LNT: Much longer than that. We have a fence right here of another species that Johan got for us. Posts and rails for fencing, and its specific gravity of about 1.1 or 1.15. I asked him if it was durable and he said, you remember when you sent me way back to check on some land corners. I said yep, and then he said well, I found a corner post back there in 1972 that had a scribe. He said it had a scribe date of 1902 on it. It was just as solid as it could be right now.

HKS: They were in the process of clearing the land to establish plantations. The pulpmill's way in the future.

LNT: Way in the future.

HKS: And they're still planting gmelina?

LNT: They're still planting gmelina.

HKS: Then Clayton does his magic act with Ludwig and shows him the pines.

LNT: He had to do it on the sly. When I went down the first time they were very busy clearing land, and they already had the nursery in place and they were promptly replanting all the areas that they had cleared with this burning. I mean, within months all of them were replanted and revegetated. They were very careful not to leave the soils exposed very long. Gmelina would just take off and provide cover very quickly.

HKS: From John Welker I got the impression that the strategic plan changed through time. They were going to use gmelina for lumber and plywood.

LNT: Yes.

HKS: It wouldn't grow straight. I was trying to figure out how they knew how many acres to clear, how much plantation they needed, and how much and how fast it grew, the volume production per year to run a pulpmill or a sawmill. I mean, it's kind of complicated to figure that out.

LNT: Yes it is. Ludwig had discovered gmelina; originally it's indigenous to southeast Asia. I early on began to caution Clayton and Johan that with relatively poor form—a lot more crooks and branches—they were going to find that harvesting cost was going to be a good bit higher, and it was.

HKS: Did they begin right there at the town and start clearing land, or did they go out and select a site and follow some sort of a topographic line or something, because they wanted good sites to grow gmelina.

LNT: Right.

HKS: Whatever a good site is. I'm not sure if a good site for gmelina is a good site for pine.

LNT: Much of that land was a good bit more sandy and they were planting gmelina, early on, on those sandier soils and not getting outstanding results. That's when Clayton got to working on the pine business, and he did, in my judgment, a fabulous job of keeping up with the seed source on everything they took down there and determining that the mountain variety of *caribaea* from Honduras was the best. He tried loblolly. He tried slash pine. Both of which have done extremely well in southern Brazil, temperate zone, but didn't do well. *Caribaea* did well. Once Ludwig accepted the fact that pine could fit into that picture, they began to more and more plant pine on the poorer sites. They also had a crew. A Brazilian forester named Everaldo headed that up and they'd be out fifteen, twenty miles at times, I think from any real road or anything else doing soil surveys and so on. It was pretty well scheduled and planned, particularly after the first two or three years. It was a tremendous rush to try to get the thing off the ground in a hurry.

HKS: Sure.

LNT: But they began to do it lot more carefully after the first few years of experience there.

HKS: I've seen some aerial photographs of the plantations. I don't know how typical they are, but it looks like they are following some kind of a contour line up a valley. Like they made a decision where to stop, and somebody was putting some kind of markers at places. I don't know if you were involved in judging the logic of their location where they logged or not.

LNT: I really wasn't. I'm not a real good soil scientist.

HKS: So it wasn't obvious by what was growing on it when you saw certain native species that this would be a good site.

LNT: There are good soil indicators down there. I can tell you one tree that is indicative of good soil down there is the one on which the Brazil nut grows. The Portuguese name is *castania*. I've forgotten the Latin name for it. They were always good indicators of better soils and there were several other plant indicators. But as I recall they did some pretty darn thorough soil testing in setting up what they planted after that because obviously they had roughly three million acres there to work with, and they knew they weren't going to try to clear three million acres.

HKS: They took something like ten percent in total.

LNT: Right. Of course, they had to look at transporting and all of that. They really did not want to strip enormous areas, which is one reason you see from the photographs a good many strips of native forest left there.

HKS: I want to dwell on this a little bit because of the environmentalist criticism.

LNT: Right.

HKS: Was this a practical thing or a political decision not to cut it all?

LNT: I'll say very quickly, no it wasn't. They were always sensitive to the fact that they were the biggest project in the Amazon since Ford tried to establish Fordlandia, and it didn't work for the rubber. But I would just give them credit for being pretty sensitive to what really made good environmental sense.

HKS: Ludwig, who might not have been attuned to these policies, was not critical of these kinds of on-the-ground decisions? He might say, as long as you're there, why don't you just take it all?

LNT: As far as I know, he gave them a lot of latitude there. He sure did. Now he at times could be a dictatorial guy and he was always in a hurry, understandably. Clayton, Johan, and John deserve a lot of credit for not going out and making some major environmental mistakes. I think any really objective person who went to the project today and looked it over would concur with what I've said.

HKS: It's easy to look back and see mistakes. It seems strange that in the very early years before you went down they didn't have enough seedlings in the nursery to plant the land they cleared. It doesn't seem like it should have been a surprise that they were clearing so many thousand acres and it takes so many seedlings per acre.

LNT: Yes.

HKS: And yet the nursery seemed to be out of sync with the land clearing, or is it harder than it seems to make that kind of coordination?

LNT: I think it probably came about when they didn't really have their organization meshed and maybe a guy may have been in charge of the nursery and maybe one group not knowing what the other one did.

HKS: It felt like that G.P. operation in Florida.

LNT: Yes. Yes. I suspect that may have been the background. They had to learn how to grow those gmelina seedlings where they would survive when out-planted. Clayton came up with that system of chopping them off and planting stumps so to speak—which worked. So I think there was some learning curve there as well.

HKS: I have in the outline that one of the things you consulted on was the plan and design facilities in synchronization with the market forecast. Is that a correct assessment? It doesn't sound like you really were doing that from what you just said.

LNT: Certainly not in the early days. We got into a number of discussions and I even made a trip to Everett, Washington, the Weyerhaeuser plant where Mr. Ludwig had shipped some gmelina for manufacturing the lumber. It was pretty interesting. So small gmelina logs about that big [gestures] on one of the old big Weyerhaeuser sawmills.

HKS: I went to school in Seattle and I took a tour of that mill. They put some pretty big logs through there.

LNT: It was pretty funny.

HKS: Weren't there better mills to test it out? Hardwood mills here in Georgia, for example?

LNT: Yes. I don't know how it worked out. He and George Weyerhaeuser even considered pretty seriously a merger or Weyerhaeuser taking an interest in Jari. At any rate, I participated fairly early in the value that I thought they might receive out of lumber and possibly plywood.

HKS: So you were consulting directly with Clayton and Johan then, not with the people in Stamford, Connecticut.

LNT: Yes.

HKS: They just wanted your advice, your reaction to what they were doing.

LNT: Yes.

HKS: How long would you stay at Jari?

LNT: Typically I'd go down and stay anywhere from a week to maybe a max of two weeks. Primarily because I had a small business going here and I couldn't spare much more time.

HKS: Where did you stay at first? What was the housing like? Pretty primitive or was it well along by the time you got there.

LNT: It was well along by the time I got there. Most of the time I stayed with Ann and Johan. I was that close to them. They had a decent home there and the food was fine. They had a water purification plant, unlike Portel where I brushed my teeth with bottled water. Later a very nice guest house was built at Monte Dourado, and I stayed there.

HKS: Welker described the standard house: two bedroom, one bath, kitchen, living room, and sort of t-shaped. Did everyone stay in that same basic house or was there a hierarchy so that Johan had a bigger house than John?

LNT: I don't remember. Probably there was a little bit of hierarchy there, but they ran a pretty egalitarian set up among the professionals down there.

HKS: I guess Bob Gilvary designed those houses.

LNT: I think he did. He's a very resourceful guy. He, everything considered top to bottom, might have made the greatest single contribution to the Jari project. He's a real sound engineer. He can get things done building roads, building bridges, building whatever, and he worked marvelously well with the Brazilians. I'll put it this way, if they hadn't had Bob Gilvary or a guy with comparable skills, it would have been a lot more difficult for sure.

HKS: I can imagine. My limited forestry experience is all in the Pacific Northwest, but similar in its development of an old growth situation. I was a lot closer to civilization, but still it's the same. You have to put the roads in the right place. They can't wash out.

LNT: Yes. Bob Gilvary was a key figure, I'll put it that way, in getting the job done there.

HKS: Logging the native forest; that was essentially clearing, but there was a little for the hardwood sawmill for local use.

LNT: They would set up areas that had significant potential for lumber as far as the species and the size of timber.

HKS: Some of the species were known well enough that you knew they would make a certain kind of lumber, but others you didn't know anything about?

LNT: Right. They would go in ahead of any clearing and put in a reasonable road and you could log most of that timber with conventional equipment. Mainly they used tractors, front-end type loaders. They used a good bit of this wood, *castania* grows as a medium texture where it wasn't so dense. It made decent lumber for housing. There could have been possibly some better utilization of a portion of that native forest.

HKS: The stuff doesn't float, a lot of it.

LNT: That's right.

HKS: You're a long ways from a manufacturer. You would have to have a mill there or ship. Where would you ship them if you shipped them down river. Does Belém have a sawmill capacity?

LNT: Not a great deal. They ship some of the floaters.

# The Big Sawmill

HKS: From Jari obviously you could have shipped it out to the world market. Clayton said that one of his biggest failures was the building of that big mill, right out of Everett, Washington.

LNT: I pushed somewhat for looking more closely at selected native species that could be processed, you know, into the lumber that could go into the world markets. It was finally done much later with that great big sawmill, and I'll tell you the story on it. They had a lot of things they needed to be doing. In context of the whole big picture, doing what I was talking about doing didn't have a real high priority and would have caused some distraction. They had some small mills there, Mighty Mites and so on. They could get their construction lumber and stuff cut. So as a consequence, for a long time there was nothing done about it. There was some discussion about the big sawmill. It was coming along pretty much simultaneously with the pulpmill. This was like in 1978 and at that time H. A. Simons Company in Vancouver had been engaged to do the onshore facilities for the pulpmill. The pulpmill itself and the power house were built in Japan. I'm sure that you're aware of that.

HKS: Right.

LNT: Towed around Africa up to Jari. I wish I'd been there when it came in. I was there before and I was there after it was set in place. Interestingly enough, a little bit ahead of this Ludwig was interested in developing a bauxite deposit that he had acquired mining rights on further up the Amazon. I don't remember how far, but like several hundred miles, and he jumped me about coming up with something where he could use waste wood or wood that was up there to fire a steam plant to make electric power and stuff they had to do. I ran down the expertise here in the U.S. for what you need and what your energy yields would be and so on. We were able to lay out a pretty matched up project for that. Nothing as big as the pulpmill. He never did go ahead with it, but he called me one day and says look, you remember when I had you working on this thing about the wood fired plant for the bauxite. I said yes sir. He said he been burning up a lot of forest down there in our clearing. Why don't we go with a wood fired boiler plant for the pulpmill and save on energy cost? Brazil has very little oil reserves and their only coal is down in the south and it's more or less a brown coal. It's not a high energy coal like we get. So there was some considerable savings possible. I said it sounds awfully good to me.

HKS: And you studied it for the bauxite plant. You knew how many tons or cubic feet it took to generate a kilowatt.

LNT: Yes sir.

HKS: There was ample wood on the ground to generate kilowatts to run a pulpmill.

LNT: Right, and it was going to waste. So, we were going to build a sawmill and a facility to generate the fuel chips. I argued back and forth. Separate the facilities, I said, otherwise when you put a sawmill in there all you're going to be doing is mostly making chips. Ludwig had talked about Henry Byrd who did the engineering on the mill here and did the engineering on a bunch of G.P.'s hardwood mills for me and had been down there with me one time. Ludwig said I'd really like to get you guys to do this thing, but anyway it turned out they decided to use Simons.

HKS: Back to this energy. Maybe it's not practical to take too long a time frame, but at some point you're no longer clearing land.

LNT: That's right.

HKS: At some point you're no longer putting in plantations. Where does the energy come from when you reach that stage? Or was that an issue you worried about at that time?

LNT: I can't tell you precisely what the energy balance is down there now. There's areas in the tropics where it's a practical matter to grow cellulose for fuel, particularly in the areas where your traditional sources of energy are quite expensive. Oil, or coal, or what have you.

HKS: So you have plantations that "produce energy". It's a feasible thing?

LNT: It's a feasible thing. You also get a good bit of energy out of the pulpmaking process where you recover the black liquor. I've forgotten what they call it. It's a pretty damn tricky process but they all do it. They recover a good bit of energy, basically, out of the lignin that's discharged in the process. You also get energy out of the bark, any kind of waste.

But, when they decided to go with Simons that was fine. I made a couple of trips to Vancouver during design with the Simons people and was able to make some suggestions that helped, but it got to be pretty frustrating. I'd go out there and spend several hours or whatever going over the plans. They said, okay, we'll do it. We'll send you prints and modify it. When I got back those prints had come in, and they often weren't changed. Finally I called Ludwig up. I said, look, Mr. Ludwig, I'm wasting my time and your money. Count me out of this deal. He said, okay, Tommy. I understand where you're coming from. They built it. Looked like a battleship and soon as they started running it, the phone started running off the wall. It was almost scaled in like what you're talking about the big mill at Everett.

HKS: The mill could be built on site out of local materials, but yet you bring in the hardware, the saws, and the carriages, and, but all the beams and all the rest of it is local material.

LNT: That's another good story.

HKS: Okay.

LNT: That's what I recommended because we had done it at Portel. These dense durable woods you have down there the people are still used to working with timbers with foot adze and fitting and all of that, and I strongly suggested that they use native timber to build the structure of the new sawmill.

HKS: How were these decisions based? Is this part of Ludwig's influence.

LNT: No, I'm going to lay a little bit of that on Clayton's shoulders now, because Clayton was the guy to whom I suggested doing that. "No, we can't do it." Some insurance cost or something. Ludwig called me one afternoon livid. Said I just saw where we have purchased eleven hundred tons of steel to build that sawmill, and when Ludwig was really upset he could cuss a blue streak and not repeat himself for an hour. That was the angriest I had ever seen him. I didn't tell him what had happened, because I figured he'd fire Clayton and everybody else involved, and I think he would have that day. Anyway, it was built on concrete and steel.

HKS: What was the capital investment then?

LNT: Eight million bucks. Eight million bucks. Later when I was down there and saw Ludwig face to face, I said, "You remember that day you called me raising hell about all that steel?" He said yes, and I noticed he looked a little sheepish. I said I didn't want to tell you that day because I figured you'd fire half of your folks, but I had suggested precisely what you talked about. He sort of chuckled. He says those guys admitted to me later that you had suggested doing that. You know, it didn't take any brilliance or anything else to conclude that where you've got that class of structural timber that's got infinite life to it, and you got people used to working with it. The other thing, Clayton said initially was, well, we're worried about fire. But they built that concrete and steel plant and put in an excellent sprinkler system. That was all you needed for the wood. What happened was exactly what I was afraid of, that getting the chips produced for the boiler plant took priority. The facilities for cutting lumber weren't anything like what they should have been. You know, it was impractical and so on. So it ended up really not being a heck of a lot more than an expensive chip mill. It didn't have to happen that way.

HKS: When you look at a tropical forest and you see some of those big butt logs, I can see why you would want a big head rig, but are those the trees that were milled? The big ones with the fluted trunks?

LNT: No, you can't really do that much, you've got to get out of that fluting.

HKS: Well, how big were the logs?

## Need to Sort Logs

LNT: There were three- and four-foot diameter logs, but mostly those were the species that were being broken down and put into that fuelwood. Particularly something like *Lignum vitae* there with a specific gravity of 1.1 or something. A lot more energy there per cubic volume. Another thing that I tried to convince them on and failed was to separate the species that had real potential for lumber in the woods. Bring them in separately and so on. They had also asked me to send my son Larry down there on some technical aspects of saw filing and lumber manufacturing and all of that. And Larry was aware of that. He was in a meeting, and the logging supervisor just pitched a damn fit and says, oh, hell. There ain't no way we can do that with logging. Well, I knew damn well better and Larry did too, but Larry kept his cool.

HKS: Sorting logs in the woods is not an exotic process. Throughout the States it's routine.

LNT: Particularly today in our harvesting we are sorting at least five to six sorts with different species going to different places. Then cordwood going to chip mills and all that.

You get occasionally very valuable species; it comes to mind first the wood called sucupira that's very valuable in Spain and places like that. I mean a dollar a board foot or more. You wouldn't even had to saw it; just buck the logs out and accumulate them and you'd have got the dollar a board foot for them. They'd wind up going in the chip pile. It was never really meshed together as it was intended, but the equipment was not ideal either.

HKS: I suppose part of it was the focus on pulp for the mill. Everything else you could rationalize as "wasteful" because the pulpmill was the big thing, but it does seem strange. What little I know about Ludwig, he looked for opportunities to maximize returns.

LNT: Yes.

HKS: You grow rice here and you grow cattle here and you grow trees there.

LNT: I would under no circumstances consider hiring H. A. Simons as an engineering firm, period.

## Need for a Railroad

LNT: I think a little bit of input from me on the railroad would be worthwhile. This occurred let's say around 1980. We were getting a bit further along toward really operating, rather than just planting. By then I was hearing from Ludwig directly, occasionally requesting me to undertake certain things. He called one day and said I want you to go down to Jari and do something special for me. I said, "Well, what's that." He said I want you to go down and convince Johan and Clayton that we need a railroad. By then I had had enough experience with him that when he came up with what seemed like sometime a way-out idea it was a good idea to look pretty closely at it, because he might be just that far ahead of you. In truth that was the exact situation in regard to the railroad. As I recall the industrial site was chosen because the river began to get a little shallower. That may not have been the key call there and I did not participate in that decision.

HKS: I think that's what Gilvary said. You could only go on major ships that far up Jari River.

LNT: As a consequence, the way the land lay and everything else all the wood going into that plant was going to have to go down a twelve-mile road. Once they really got into moving, it was going to be almost too much plus the travel of workers going back and forth and all of that to really use that road. At that time Clayton and Johan were figuring on bigger and bigger trucks and this and that and the other. So I load up and go down there. I was always well received by Clayton and Johan, so I was very up front.

I said now look, Mr. Ludwig asked me to come down here and convince you guys that we need a railroad, and the initial reaction was, "Naw, we don't, Tommy, you're wasting your time." I said, "No. Let's look at it a little closer now." I said Mr. Ludwig once or twice faked me out when I didn't analyze what he was thinking about. We really began to work it through just what was going to be required and reached the conclusion that as a practical matter the railroad was going to be far more serviceable than the road and more economical to operate. Capital cost of building it to start with was pretty steep.

The safety element alone. I'm not a scary cat at all. In fact, trained as a Navy carrier pilot when I was eighteen, nineteen years old. There are no cowards as carrier pilots. But, one of the most anxious days I ever had was coming back from the industrial site. It was raining cats and dogs and 631 Cat scrapers coming down the road loaded, running thirty miles an hour. Autos ducking in and out. I was cautious almost beyond belief. They had a number of fatalities on that twelve miles of road. They finally began to provide some passenger transport on the railroad. Now whether it's still being done now or not I don't know, but they did that. Anyway, when Johan, Clayton, and I really analyzed what was going to have to move over that road, we reached the conclusion that the railroad was the best way to do it and it was buildable. Bob Gilvary may have done the survey on the thing. I don't know.

HKS: He described a bit some of the technical issues. I forgot to ask him how he got the locomotive off the barge. It takes a pretty good port facility to lift something like a locomotive. He explained how easy it was to walk a Cat off. But a locomotive, you have to pick it up I understand.

LNT: Knowing Bob he probably built him a short piece of track and tied it together where he could run it off of there. This is a total aside, unrelated to Jari. In the early '50s we were shipping railroad car loads of lumber out of Savannah to Puerto Rico in cars. It was a rail car ferry. When it got to Puerto Rico they'd take them off. That's before the trucking business and the containers got called in. They had that set up very well. I had sort of forgotten about that.

HKS: I was just thinking of the salt water and the tide. It makes it a little harder to imagine how those tracks link up, but I guess they worked it out.

LNT: There's a little bit of tide at Jari even that far from the ocean. As I remember it, it's two feet. Something like that. But there's a significant one.

HKS: But the railroad is one of the things that the journalists use as an example of Ludwig's vanity. That he had to have a railroad just like a toy. However, the railroad seems to make wonderful sense.

LNT: If I were going to tackle any major movement of freight, particularly in the tropics, the maintenance of roadways when you're getting a hundred inches of rain a year and more in hard downpours is no small undertaking. I would look long and hard before I'd pass up the railroad option.

HKS: This is one of those "ifs", if they had known initially that there was going to be a railroad with the less flexible road alignments, would the overall design of the operation have been different?

LNT: It was getting well along in the game. They would probably have been some changes. Yes, I think. But, still it fit a genuine need and the safety of that thing I'm not kidding you. Now if you'd have put all the trucks on that roadway in addition to what was already there, I will describe it as an impossible situation in terms of safety alone.

HKS: Did the train haul a load a day or make two trips a day?

LNT: I can't tell you on that. I'm just not sure. It made at least one trip a day, as I remember, pulling ten cars or fifteen cars of wood. But they probably might have been pulling two a day. I'm just guessing now.

HKS: I really can't visualize how many truck loads fit on a railroad car.

LNT: Roughly three to four truck loads fit on a rail car. So they probably were making two trips a day.

HKS: Gilvary said that basically the railroad cars are all custom made. That's the length primarily.

LNT: Yes.

HKS: It depends upon the radius of the curves it has to navigate. I guess it's common when you buy a railroad car you tell them how long you want it to be. There's no standard length like it is in the States.

LNT: Well, no. There's more and more variation. We have Norfolk Southern service here on all three of our plants strangely enough, and some of those cars today are getting up to seventy feet. I'm sure there's probably some track that they may not put them on. They may have tight enough curves they can't handle it.

HKS: Were there two locomotives or four?

LNT: I think it was probably two.

HKS: We've already talked a bit about using waste wood from land clearing for power plant fuel. Is there more to say on that?

## Ludwig the Man

LNT: There may be more of a story. It was working well about the time that Ludwig decided to sell out. I know I had a visit with him a couple of hours, just the two of us, and that's the last time I saw him. I kept promising myself to fly up to New York and see him one day, and never did it.

HKS: It's probably a good time to talk about Ludwig the man. He's described in various ways. One article described him as being very tall and yet I've seen a photograph of him standing with Ronald Reagan, and Reagan is half a foot taller. What's your impression of the man?

LNT: Ludwig was no taller than me. He might have been an inch shorter.

HKS: For the record your height is?

LNT: Five-ten.

HKS: He's not unusually big.

LNT: No.

HKS: Maybe it was a small journalist who wrote the story. [laughter]

LNT: He was of average build. He probably most of his life weighed a hundred and sixty pounds or something like that. Had a pretty serious back injury early in his life in a shipwreck saving the lives of two of his seamen. It plagued him pretty much the rest of his life. I asked him about it one time and he said, well, says we keep talking about an operation and said the doctors let me have so many aspirins a day and so on. He eventually did have an operation on his back before he died. He was to me one of the most original thinkers that I've ever met.

HKS: You, Clayton, Bob, and John have certainly one thing in common. You all call him mister. It must have been something about his bearing.

LNT: Oh, yes.

HKS: You wouldn't think about calling him Dan or D. K.

LNT: It's not just us. You know who Locke Craig was?

HKS: No.

LNT: Okay. Locke's father was governor of North Carolina way back, and Locke graduated from N.C. State in the '30s as a forester. He worked with forestry service there in Asheville for awhile then spent some time in the Belém area with Jari. In the very early '50s he was working on a research project when he was down in that area of Brazil. Went out to Indonesia and ran a big rubber plantation for, it was either Goodyear or Firestone, I've forgotten which, for maybe as much as five or six years. Came back and went to work with one of the companies in southern Brazil, an American company, Olin Kraft. He took over just a little pepper box papermill down there and built it into a damn first class operation. In about fifteen years he built up a land base. He retired from that company and went on Ludwig's Jari board. Locke died about three years ago down at Southern Pines where he had retired and he was pushing eighty. A really capable guy and a real figure. He and I were talking one day and he said one time that Mr. Ludwig told him, "Locke, why don't you call me D. K.?" He said, "I'm sorry Mr. Ludwig, but you're Mr. Ludwig to me. I'll just have to keep calling you that."

All I'm saying is it wasn't just us younger guys that felt that way about him. I'm sure he had contemporaries that may have called him on a more personal basis. The newspaper reporters who thought he built that railroad as a ego question didn't know the man. The man was in no way interested in the praise and fame from other people. I don't know whether you ever heard the story. He really provided the spark and the money to kick off the Japanese shipbuilding industry after World War II, and for a good many years built his own ships in a yard at Kobe. The Japanese people wished to give him the Order of the Chrysanthemum, which is the highest civilian honor in Japan. He said no, I'd rather you give it to the man that manages the shipyard. He most of the time, I'm sure someone's told you this, flew tourist class.

HKS: Sure.

LNT: I was in New York one time for a meeting with him and some of his people. He walked from wherever his apartment was to his office there on the Avenue of Americas. He was also quite careful to try to avoid being photographed for the simple reason he traveled by himself a great deal worldwide, and you know a man worth the kind of money he was in his heyday is a mighty tempting kidnapping target. But as I say, he was just the kind of guy that did not try to parade off with a lot of pomp and ceremony. But he was tough as pig iron to work for. I might not have gotten along as well with him as an employee as I did as a consultant. I don't know.

HKS: How did you get started with him? I mean, was it logical he knew you were down there at Clayton's invitation or was there another way that he knew about you?

LNT: No. It would have had to come from them, because he wouldn't have known any other way. It's really interesting. He had a great deal of respect for a guy that ran his own business like I did.

HKS: Okay. I can understand that.

LNT: He wanted me down there more than I needed to be away from here. Not being a smart aleck I'd say, "Mr. Ludwig, if I leave I'm going to be neglecting something here in my own business." He'd say, well, let me know when you can come. He had a tremendous regard for individual entrepreneurs.

HKS: He didn't think a whole lot of college professor types.

LNT: No.

HKS: Theorists. Academics. Although Clayton has a Ph.D., and they seemed to get along.

LNT: Oh, yes. He took an advanced course at MIT in marine engineering, I think. You know you don't get in advanced courses at MIT being a dummy and this was when he was in his 50s or 60s, Clayton told me. I think he earned a B+ in it [laughter] after being out of academia that long.

HKS: Did you from your observation see—I don't know how to characterize it—his impatience with personnel where he fired people, and all the turnover. You wonder about the selection process if he hired so many people who didn't work out, if that's really a fair thing to say. Apparently it's true if he fired thirty-four managers, or some very large number. He picked the managers it seems like. Why would he have misjudged people that way?

LNT: I will give you my theory on it. I've thought a little bit about that same question. Number one, this was his very first venture into a field of this nature. He'd been in mining. Of course, the transport of bulk commodities was his strength and he was preeminent in that area. No question. But he was in a somewhat different ball game, and he really did not have people in-house, in his own organization in New York, who had any background in what was going on. And he was in a big hurry. Throughout the time that I knew him or knew what went on before I got involved with him, he wanted to see it all done before he left the scene. As I recall he started when he was sixty-nine. My opinion is that he probably got some poor input on these people. Obviously, somebody made some poor decisions there. I knew all this was going on when I was involved at Portel, and I'd meet the occasional manager coming or going.

HKS: I can understand how he would misjudge the complexity of the biology of growing things as opposed to mining things.

LNT: Right. Right.

HKS: But you don't become a billionaire by not being a good judge of character.

LNT: Right, I've got to tell you one other story. One other project he put me on. You were asking about sawmills and so on. He calls me up one day, here, and says I want you to figure out how to put a sawmill on a ship. I said, Mr. Ludwig. I'd gotten to know him pretty good then. I said, I don't know a damn thing about ships. I know they float. I can't do you any good on that. He says I know you don't know a damn thing about ships, but he says, I'm going to send you a blueprint of this thing and you and Henry Byrd get together and figure out how to put a sawmill in this thing, and we'll put all the equipment in it here and have it ready to go. He had a liberty ship carried over from World War II. And we'll move it down there. I want to put a sawmill in there. We'll have all the machinery on the ship to operate the sawmill at Munguba, the industrial site. I said okay. So I got with Byrd and we worked and worked. Finally figured out a way to put the sawmill in the liberty ship and made it run, but the biggest trouble was waste disposal, sawdust and all of that. We never did do it. I was real glad when he finally decided to forget that idea.

HKS: But that's part of his philosophy on the pulpmills. You're going into a Third World situation. You barge them in and you don't have to build an infrastructure there. It's a way of developing.

LNT: That idea of his about putting the pulpmill and power house together in Japan was, in my judgment, brilliant. When you think about trying to build that complex of entity piece by piece in a remote area like that and house the technicians to do all of that installation, it would have been enormously more expensive. All of the knowledgeable paper people (and I'm not knowledgeable at all about paper) told me the Japanese did a superb job with the pulpmill and power house. I bet you he saved a hundred million dollars minimum on the cost of the plant.

HKS: Clayton said that was an idea he developed with this friend in Crown Zellerbach to be tried in Honduras. This is before the Jari project really materialized and that concept had been in his head for awhile.

LNT: I didn't know that that concept had come up earlier.

HKS: Did you experience this characterization of him—you worry about getting the job done and let him worry about the money. You wouldn't go ahead if it had been your own mill, but he overruled because it saves six months.

LNT: I never did get into head to head confrontation with him, but I've been asked by a number of people over the years what was Mr. Ludwig's biggest mistake in regard to Jari. And my answer has been, and this is strictly my own opinion, that the thing that caused the project to cost as much as it did was that Mr. Ludwig was in a genuine hurry, and time and again something would come up and he'd want to do it and said as he should, what's it going to cost to do it and how long's it going to take. I don't know what it might be, but they'd come back and say well, it'll take a couple of million dollars and a year and a half or two years. And he'd say, well, what's it going to take to do it sooner. There was a great deal of that done, and in my judgment even those at Jari used to the idea that you really didn't have to worry that much about what this was going to cost or what that was going to cost.

## Potential of Low-Impact Silviculture

HKS: I've seen the video tape Johan put together on the Tropical Forest Foundation. I guess the best way to characterize it is low impact silviculture, as opposed to plantation silviculture. Do you have any sense that's a viable way to provide wood commercially?

LNT: That's real interesting that you asked me that because in my early days with Mr. Zweede, he was convinced that the only forestry that was worth a damn was plantation forestry. I kept telling him I feel that you can manage these natural stands and the Amazon on a long-term basis and do well at it. So much of European forestry is plantation forestry whether it's softwood, hardwood, or anything else. I said that's not right, the government of Brazil and the forestry people should be doing some extensive silviculture research now because I think it will work. So Mr. Zweede, I'm tempted to kid him a little bit and say it took him twenty-five years to figure out what I was trying to tell him, because it did work and there was no question.

HKS: Okay. The question in my mind for low impact silviculture would be the higher cost of road maintenance for its relatively lower volume per year over those roads.

LNT: As long as you're not trying to get in the year-round logging, Pete. See you get most of the rainfall down there in seven months. There is some rain at other times. I don't mean it never rains, but you don't get the heavy rains. As long as you don't have to go try to volume log during the rainy season you don't have to spend so much money on roads and you don't mess up your roads if you can do that. If I were going to try to operate down there, I'd approach it this way. Probably do the same thing we are about to do here and we already have put in water spray storage for the wet months so that you can keep operating. It's getting economically almost impossible to operate a sawmill on seven or eight months out of the year and shut it down for weather the other time. I mean you count your people cost and everything else just leaves you flat. But I think it's a practical matter and I tried to tell that hardheaded Dutchman twenty-five years ago that it'd work. Plantation forestry has worked at Jari, but the Amazon basin contains literally millions and millions of acres of this very diverse hardwood forest.

HKS: That's a good note to end on.

# Further Reading

Blount, Jeb. "Brazil Tree Farm Uses Rain Forest and Also Saves It." *Christian Science Monitor*, 17 April 1993.

Carrere, Ricardo. "Pulping the South: Brazil's Pulp and Paper Plantations." *Ecologist* 26, no. 5 (September-October 1996): 206-214.

Fearnside, Philip M., and Judy M. Rankin. "Jari and Development in the Brazilian Amazon." *Interciencia* 5, no. 3 (May-June 1980): 146-156.

Fearnside, Philip M., and Judy M. Rankin. "Jari Revisited: Changes and the Outlook for Sustainability in Amazonia's Largest Silvicultural Estate." *Interciencia* 10, no. 3 (May-June 1985): 121-129.

Fearnside, Philip M. "Jari at Age 19: Lessons for Brazil's Silvicultural Plans at Carajas." *Interciencia* 13, no. 1 (January-February 1988): 12-24.

Kinkead, Gwen. "Trouble in D. K. Ludwig's Jungle." *Fortune* (April 20, 1981): 102-114, 117.

McIntyre, Loren. "Jari: A Billion Dollar Gamble." *National Geographic* (May 1980).

McNabb, Kenneth, Joao Borges, and John Welker. "Jari at 25: An Investment in the Amazon." *Journal of Forestry* (February 1994): 21-26.

Shields, Jerry. *The Invisible Billionaire: Daniel Ludwig*. Boston: Houghton Mifflin Company, 1986. See especially chapter eighteen, "Welcome to Brazil, Mr. Ludwig."

Welker, John C. "Site Preparation and Regeneration in the Lowland Humid Tropics: Industrial Plantation Experiences in Northern Brazil." Paper presented at conference on Management of Forests of Tropical America: Prospects and Technologies, San Juan, Puerto Rico, September 21-26, 1986.